Alexey Feofanov

Experiments on flux qubits with π-shifters

Experimental Condensed Matter Physics
Band 1

Herausgeber
Physikalisches Institut
Prof. Dr. Hilbert von Löhneysen
Prof. Dr. Alexey Ustinov
Prof. Dr. Georg Weiß
Prof. Dr. Wulf Wulfhekel

Experiments on flux qubits with π-shifters

by
Alexey Feofanov

KIT Scientific Publishing

Dissertation, Karlsruher Institut für Technologie
Fakultät für Physik
Tag der mündlichen Prüfung: 03.12.2010
Referenten: Prof. Dr. Alexey Ustinov, Prof. Dr. Alexander Shnirman

Impressum

Karlsruher Institut für Technologie (KIT)
KIT Scientific Publishing
Straße am Forum 2
D-76131 Karlsruhe
www.ksp.kit.edu

KIT – Universität des Landes Baden-Württemberg und nationales
Forschungszentrum in der Helmholtz-Gemeinschaft

KIT Scientific Publishing 2011
Print on Demand

ISSN 2191-9925
ISBN 978-3-86644-644-1

Experiments on flux qubits with π-shifters

Zur Erlangung des akademischen Grades eines
DOKTORS DER NATURWISSENSCHAFTEN
von der Fakultät für Physik des
Karlsruher Institut für Technologie (KIT)

genehmigte

DISSERTATION

von

Magister für Angewandte Physik Alexey Feofanov
aus Zaporozhye

Tag der mündlichen Prüfung: 03.12.2010

Referent: Prof. Dr. Alexey Ustinov

Koreferent: Prof. Dr. Alexander Shnirman

Contents

Chapter 1

Introduction

1.1 Quantum mechanics

What is quantum mechanics? Quantum mechanics is a set of rules for the construction of physical theories that describe the behaviour of the microscopic world. The rules of quantum mechanics are simple but even experts find them counterintuitive, and among the earliest motivations of thinking about quantum computing and quantum information, one may find the long-standing desire of physicists to better understand quantum mechanics.

The best known critic of quantum mechanics, Albert Einstein, helped to invent it. In 1905, he published the paper "On a heuristic viewpoint concerning the production and transformation of light" [1], where he proposed the modern concept of the photon, a quantum of the electromagnetic energy. This idea was motivated by an earlier work on black body radiation by Max Planck [2]. The theory of light quanta was a strong indicator of wave-particle duality, a fundamental principle of quantum mechanics. But thirty years later he tried to disprove quantum theory. In 1935, Einstein, Podolsky and Rosen introduced a "Gedankenexperiment" that was intended to show that quantum mechanics cannot be a complete theory [3]. They argued that measurements performed on spatially separated parts of a quantum system can apparently have an instantaneous influence on one another. Einstein considered these nonlocal interactions to be impossible and called this effect "spooky action at a distance". In quantum theory, this phenomenon is known as quantum entanglement. In 1982, Alain Aspect published a paper detailing his experiments showing that nonlocal interactions do occur [4].

1.2 Quantum computing

In 1985, David Deutsch tried to define a computational device that would be capable to efficiently simulate an arbitrary physical system [5]. Because the laws of physics are ultimately quantum mechanical, Deutsch considered computing devices based upon the principles of quantum mechanics. These devices, quantum analogues of the machines defined half a century earlier by Turing [6], led to the modern conception of a quantum computer.

The recent excitement in this area of research was triggered by Peter Shor who showed how a quantum algorithm could exponentially "speed-up" classical computation and factor large numbers into primes much more rapidly (at least in terms of the number of computational steps involved) than any known classical algorithm [7].

The implementation of Shor's algorithm on a large scale quantum computer might result in the destruction of modern electronic commerce, since widely used RSA cryptography protocols rely on the assumption that all known classical worst-case algorithms for factoring take a time which grows exponentially with the length of their input.

A classical computer has a memory made up of bits, where each bit represents either a one or a zero. A quantum computer is a set of quantum bits (qubits). A single qubit can represent a one, a zero, or, crucially, any quantum superposition of these; moreover, a pair of qubits can be in any quantum superposition of 4 states, and three qubits in any superposition of 8. In general a quantum computer with n qubits can be in an arbitrary superposition of up to 2^n different states simultaneously (this compares to a normal computer that can only be in one of these 2^n states at any one time). A quantum computer operates by manipulating those qubits with a fixed sequence of quantum logic gates. The sequence of gates to be applied is called a quantum algorithm.

There are criteria any candidate for a physical realisation of a quantum computer must satisfy. These criteria were originally formulated by David

P. DiVincenzo [8] and therefore are known as "DiVincenzo criteria".

The candidate must:

1. Be a scalable physical system with well-defined qubits;

2. Be initialisable to a pure state such as the ground state;

3. Have a universal set of quantum gates;

4. Permit qubit-specific measurements;

5. Have decoherence times (*i.e.* the time during which the system maintains quantum coherence) that are much (several orders of magnitude) longer than time required to perform a single quantum gate.

1.3 Superconducting quantum bits

The computer I am using to type this text is a solid-state device. It was produced be means of microfabrication techniques. These techniques were developed mainly to satisfy needs of the semiconductor industry, but using them we can easily scale-up almost any solid-state device. Therefore, it is natural to look for an appropriate candidate among solid-state systems, because of their scalability.

The smallest solid-state systems that can be microfabricated have size of order of several hundred nanometers and contain huge amount of atoms and electrons, which are individual quantum mechanical systems, and even larger number of degrees of freedom. A direct access to and control of each and every system is impossible. A large number of degrees of freedom can be a reason for infinitesimally short coherence times, unless we can isolate a particular quantum mechanical system from the others.

Superconductors stand out from the rest of solid-state systems, because of a single macroscopic wavefunction describing all the electrons that take part in a dissipationless charge current (supercurrent). These electrons form a so-called condensate that is a single quantum mechanical system. Moreover, this system is isolated from other quantum mechanical systems, which exist in solids, due to an energy gap that separates excitations in a superconductor from the condensate.

Thus superconducting circuits are promising candidates for a physical implementation of a quantum computer.

1.4 Flux quantisation and Josephson tunneling

The macroscopic wavefunction results in two phenomena that are vital for superconducting qubits. The first phenomenon is flux quantisation. The magnetic flux Φ in a superconducting ring cannot possess any arbitrary value, but it is quantised [9] in integer values of the flux quantum

$$\Phi_0 \equiv \frac{h}{2e} \approx 2.07 \times 10^{-15}\ \mathrm{Wb}.$$

This quantisation arises from the requirement that the macroscopic wavefunction is single valued.

The second phenomenon is Josephson tunneling [10]. A Josephson junction consists of two superconductors separated by a thin insulating barrier, the barrier thickness usually equals to a few nanometer, through which electrons that belong to the condensate can tunnel coherently. Brian Josephson showed that the dissipationless current I through the barrier is related to the phase difference $\varphi(t)$ (so-called Josephson phase) between the phases of wavefunctions in the two superconductors by the sinusoidal current-phase relation

$$I = I_c \sin\varphi\,, \tag{1.1}$$

where I_c is the critical current,*i.e.* the maximum dissipationless current that can flow through the junction. If there is a voltage V applied to the junction, the phase $\varphi(t)$ evolves in time as

$$\frac{\partial\varphi}{\partial t} = \frac{2\pi}{\Phi_0}V\,. \tag{1.2}$$

Thus, the Josephson junction is a non-linear element without energy dissipation. It has a non-linear inductance L_J determined by the critical current

I_c and the phase φ

$$L_J = \frac{\Phi_0}{2\pi I_c \cos\varphi}.$$

This non-linearity is crucial for a qubit implementation, since it enables a circuit containing Josephson junctions to form an anharmonic oscillator and thus allows for different transition frequencies in the circuit. Therefore, a superconducting qubit can be approximated as a two (or a few) level system.

There are two relevant energy scales associated with a Josephson junction. They are the Josephson coupling energy

$$E_J \equiv \frac{I_c \Phi_0}{2\pi}$$

and the charging energy

$$E_c \equiv \frac{(2e)^2}{2C},$$

where C is the intrinsic junction capacitance.

The quantum mechanical operators for the Josephson phase φ and the charge Q across the intrinsic junction capacitance are canonically conjugated:

$$[\varphi, Q] = i \cdot 2e\,.$$

If $E_J \gg E_c$, φ is well defined, and the charge Q across the capacitance has large quantum fluctuations. When $E_J \ll E_c$, Q is well defined, and φ has large quantum fluctuations.

A variety of quantum circuits has been realised, including a charge qubit [11], a phase qubit [12], the 'quantronium' [13] (a hybrid charge-phase qubit), the 'transmon' [14](which is essentially a phase qubit, but its design was derived from a charge qubit), and a flux qubit [15]. Different realisations of superconducting quantum bits have different E_J-to-E_c ratios and are thus better described in terms of either phase ($E_J \gg E_c$) or charge ($E_J \ll E_c$).

In this thesis, we focus on the flux qubit. This specific realisation of the quantum bit is characterised by a large E_J-to-E_c ratio and a small geometric inductance of the circuit. The qubit potential is very anharmonic, which allows for fast qubit manipulation. Hence, one can perform more qubit operations in the same coherence time compared with other superconducting qubits. For a review on the superconducting qubits please see Ref. [16].

1.5 The flux qubit

A flux qubit consists of a superconducting loop interrupted by three Josephson junctions (Fig. 1.1). In this device, one junction is smaller in area and thus has a smaller critical current than the other two. The junctions has a large ratio between Josephson coupling energy E_J and the charging energy E_c

$$\frac{E_J}{E_c} \gg 1\,,$$

so the phase difference φ is a good quantum variable. In this case, flux quantisation in the loop gives

$$\sum_i \varphi_i + 2\pi\frac{\Phi}{\Phi_0} = 0\,. \tag{1.3}$$

The two quantum states are anticlockwise qubit persistent supercurrent I_p circulating in the loop $|\circlearrowleft\rangle$ and clockwise supercurrent $|\circlearrowright\rangle$. This system has a double-well potential U,

$$U = \sum_i E_{Ji}(1 - \cos\varphi_i)\,. \tag{1.4}$$

We take the electrical energy as the kinetic energy and the Josephson energy as the potential energy. Combined with the flux quantisation condition (1.3), the total Josephson energy (potential energy) is given as

$$U = E_J\left(2 + \alpha - \cos\varphi_1 - \cos\varphi_2 - \alpha\cos\left(\varphi_1 + \varphi_2 + 2\pi\frac{\Phi}{\Phi_0}\right)\right)\,. \tag{1.5}$$

We can rewrite the last equation by choosing as coordinates the sum and the difference of phases, $\varphi = (\varphi_1 + \varphi_2)/2$ and $\xi = (\varphi_1 - \varphi_2)/2$:

$$U = E_J\left(2 + \alpha - 2\cos\xi\cos\varphi - \alpha\cos\left(2\varphi + 2\pi\frac{\Phi}{\Phi_0}\right)\right) \tag{1.6}$$

$$\approx E_J\left(2 + \alpha - 2\cos\varphi - \alpha\cos\left(2\varphi + 2\pi\frac{\Phi}{\Phi_0}\right)\right)\,. \tag{1.7}$$

The latter transformation is possible, since we are working in the limit of a well defined Josephson phase and thus $\cos\xi \approx 1$. Hence, the flux qubit

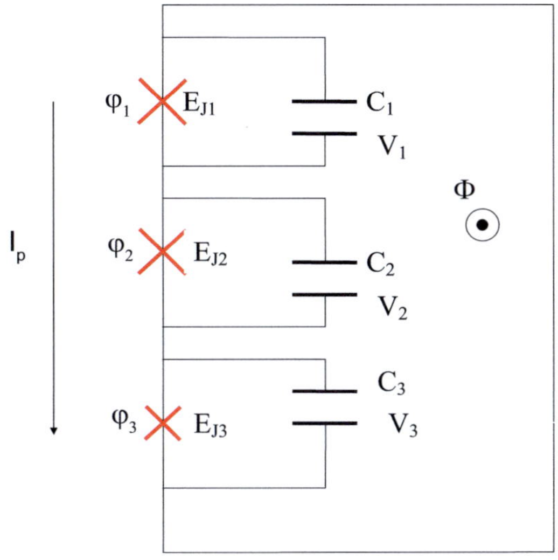

Figure 1.1: A flux qubit. The circuit consists of three Josephson junctions. Two Josephson junctions are considered to be identical, $E_{J1} = E_{J2} = E_J$, $C_1 = C_2 = C$, and $E_{J3} = \alpha E_J$, $C_3 = \alpha C$.

potential is pseudo one-dimensional. The two states are coupled by the quantum-mechanical tunneling of φ through the barrier separating the wells. When the externally applied magnetic flux $\Phi = \Phi_0/2$, the double-well potential becomes symmetrical, and the two eigenfunctions become symmetrical and antisymmetrical superpositions of the two basis states:

$$|\Psi\rangle = \frac{1}{\sqrt{2}} \left(|\circlearrowleft\rangle \pm |\circlearrowright\rangle\right) . \tag{1.8}$$

In the vicinity of this degeneracy point, the system can effectively be described as a spin-1/2 particle in a magnetic field. The Hamiltonian of this simplified two-level system is

$$H = -\frac{\varepsilon}{2}\sigma_z - \frac{\Delta}{2}\sigma_x , \tag{1.9}$$

where σ_i are the Pauli-spin matrices, $\varepsilon = 2I_p(\Phi - \Phi_0/2)$, and Δ is the splitting of the energy levels of the ground state $|0\rangle$ and the first excited state $|1\rangle$ at the degeneracy point, which exponentially depends on α. Away from the degeneracy point, the energy difference is $E = \sqrt{(\Delta^2 + \varepsilon^2)}$, (Fig. 1.2a).

The probabilities to observe the states $|\circlearrowleft\rangle$ and $|\circlearrowright\rangle$ in the ground state as a function of Φ is shown in Fig. 1.2b. At the degeneracy point, the mean supercurrent in the qubit loop is 0. Therefore, the probability to observe each state is $^1/_2$. At smaller values of Φ the probability to observe $|\circlearrowright\rangle$ is higher than one to observe $|\circlearrowleft\rangle$. In the first excited state the probabilities are reversed. Hence, we can detect the qubit state by measuring a magnetic flux produced by the qubit supercurrent I_p.

This pseudo-spin-$^1/_2$ model is applicable as long as the energy splitting between the ground and the first excited state is much smaller than the splitting between the first excited state and the higher states. The full quantum mechanical calculations of the flux qubit energy spectrum can be found in Ref. [17].

1.6 Decoherence

Decoherence limits the number of qubit operations that can be performed without error. It is caused by coupling between the qubit and its environment. In other words, decoherence is uncontrolled changes of the qubit state due to variations of parameters in the environment. These changes result in a loss of information about the qubit state, thus making well-controlled operations problematic.

For a quantitative analysis of decoherence, one can use a general framework as described by Ithier *et al.* [18]. It allows for a systematic study of the influence of noise in any normalised external parameter λ on the properties of the qubit. These parameters can represent any fluctuations, whether they originate from circuit induced or environmental noise.

Decoherence of two-level systems (spins) is described in terms of two rates (or times): the longitudinal relaxation (depolarisation) rate

$$\Gamma_1 = T_1^{-1}$$

and the transverse relaxation (dephasing) rate

$$\Gamma_2 = T_2^{-1}.$$

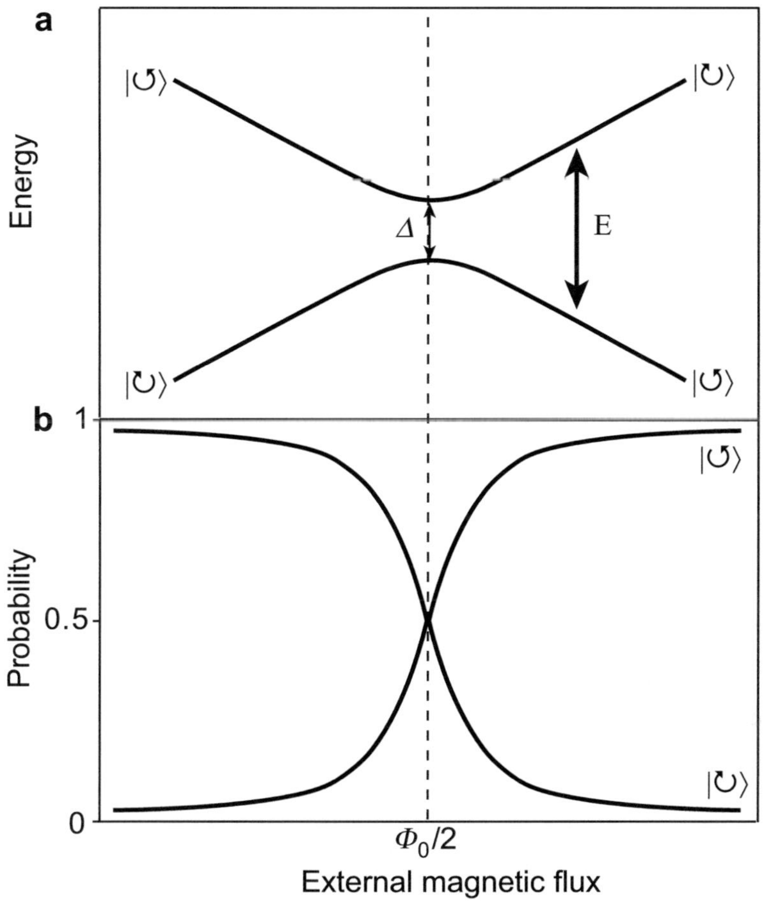

Figure 1.2: A flux qubit in the vicinity of the degeneracy point. **a**, The energy of the ground state and first excited state versus the externally applied magnetic flux is shown. The energy-level splitting is Δ at the degeneracy point, $\Phi = {}^{\Phi_0}/_2$, and is E for $\Phi \neq {}^{\Phi_0}/_2$. **b**, The probabilities of the qubit persistent supercurrent circulating counterclockwise or clockwise in the ground state versus externally applied flux are presented. In the first excited state the probabilities are reversed.

The dephasing process is a combination of effects of the depolarisation and of the so-called pure dephasing. The pure dephasing is usually associated with the inhomogeneous broadening in ensembles of spins, but occurs also for a single spin due to the longitudinal low-frequency noise. It is characterised by the rate Γ_ϕ. These two processes combine to a rate

$$\Gamma_2 = \frac{1}{2}\Gamma_1 + \Gamma_\phi. \tag{1.10}$$

Introducing the spectral density

$$S_\lambda(F) = \int \mathrm{d}t \langle \delta\lambda(0)\delta\lambda(t)\rangle e^{-i2\pi F t},$$

the relaxation rate is

$$\Gamma_1(\nu) = \pi D^2_{\lambda,\perp} S_\lambda(\nu), \tag{1.11}$$

where

$$D_{\lambda,\perp} = \frac{1}{h}\left| \langle \Psi_0 | \frac{\partial H_0}{\partial \lambda} | \Psi_1 \rangle \right|.$$

The pure dephasing time can be written as

$$\Gamma_\phi(\nu) = \pi D^2_{\lambda,Z} S_\lambda(0), \tag{1.12}$$

where

$$D_{\lambda,Z} = \frac{\partial \nu}{\partial \lambda}.$$

These expressions are linear approximations and valid when the noise is week and short correlated (*e.g.* white noise). Moreover, the spectral shape of the noise has to be smooth from $\nu \approx 0$ up to frequencies of the order of Γ_ϕ. There are alternative approaches to translate the spectral shapes to dephasing rates, which can be used in more general situations. See Ref. [18] for an in-depth discussion.

1.7 π-shifters

To realise degenerate quantum states in a flux qubit one has to supply an additional phase shift of π along the loop. This shift can be produced by external magnetic flux or alternatively by a circuit element that provides a phase difference of π between its terminals. Such circuit elements are generally named π-shifters.

In usual flux qubits, external magnetic flux is used to produce phase bias along the loop. In many experiments, an external coil with a large current and weak coupling to the qubits has been used. However, this prevents one from using a superconducting shield that provides good shielding of qubits from external flux noise. On the other hand, if one uses a local biasing with a control line, noise in the current source degrades the coherence of the qubit. Therefore, π-shifters may be an attractive option for individual phase biasing on qubits, especially in multi-qubit circuits.

There was a number of proposals put forward to implement a phase shifter in a qubit circuit. Below we briefly discuss advantages and disadvantages of some of them.

One of the conceptually easiest ideas is to replace the external magnetic field by a field produced by a permanent micromagnet [19]. A physical realisation of such a phase shifter is presented in Fig. 1.3. If properly designed, this phase shifter will provide a stable phase bias and the value of the phase difference gained along the qubit loop can be engineered to be equal to π. However, this type of phase shifters cannot be used in multi-qubit circuits with large density of the qubits because of stray fields of the permanent magnets.

Another simple phase bias was proposed by Majer *et al.* [20] It is based on a trapped flux in a superconducting ring (a flux-trap loop). If there is a flux quantum trapped in a ring, then the phase gained along this ring equals to 2π. If we include this ring in another circuit (see Fig. 1.4), then the phase difference $\delta\varphi$ over shared element will be equal to $2\pi\frac{a}{s}$, where a is the length of the shared part and s is the circumference of the ring with trapped flux. The flux-trap loop phase shifter is easy to implement, since it does not require any additional technological steps. If the length of the enclosed part equals to a half of the loop circumference, the phase bias will be reliably fixed at π, if there is a flux quantum (or an odd number of quanta) trapped in the loop. Nevertheless, trapping a well defined number of flux quanta in every

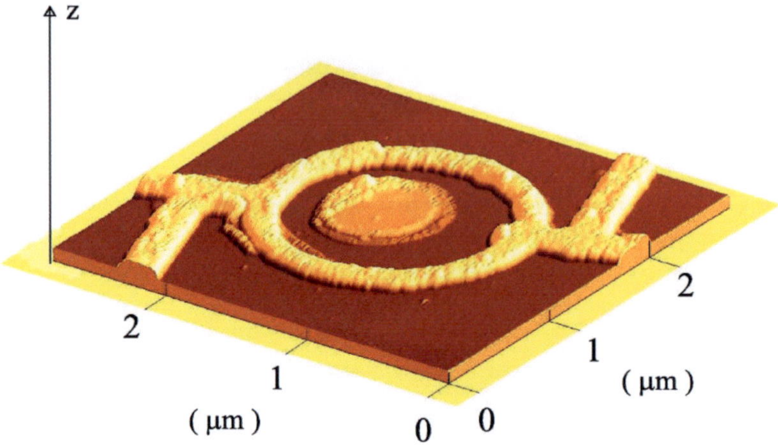

Figure 1.3: An AFM image of a physical realisation of a phase shifter based on a permanent magnet [19]. The permanent magnet is the disk inside the circular loop.

flux-trap loop in a multi-qubit circuit might be not an easy task.

To realise π-shift in a loop one can make use of a copper-oxide d-wave superconductor: making contacts to the (1,0,0) and (0,1,0) surfaces of a d-wave material (*e.g.* $YBa_2Cu_3O_7$) [21]. The condensate wave function has different signs on those surfaces that corresponds to phase shift of π. Unfortunately, it is not easy to incorporate single crystals of high-T_c superconductors in qubit

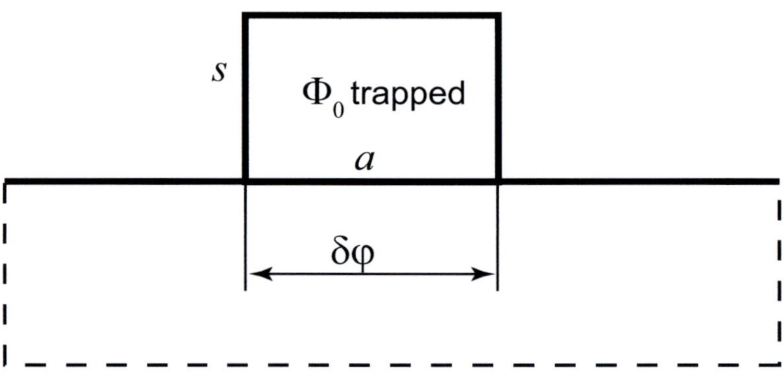

Figure 1.4: A phase shifter based on a flux-trap loop [20]. a is the length of the shared part, s is the circumference of the flux-trap loop.

circuits.

Another possibility to implement a π-shifter is to include a Josephson π-junction [22] into the qubit loop. π-junction is a Josephson junction that features a π-phase shift across in the ground state. Since this device is a Josephson junction, its phase difference is connected with the supercurrent flowing through. The current-phase relation differs from the usual one (see Fig. 1.5 for comparison) and has the following form [22]:

$$I = I_c \sin(\varphi + \pi) = -I_c \sin(\varphi)\,. \tag{1.13}$$

We note here that for a π-junction the sign of the supercurrent is reversed in respect to the sign of the phase drop over the junction. This behaviour can also be described by introducing negative critical current and negative Josephson coupling energy.

To have the phase bias provided by a π-junction close to the required value of π, the supercurrent I flowing in the qubit loop must be much smaller then the critical current of the π-junction I_c:

$$I \ll I_c\,.$$

This restriction makes, for instance, π-junction based on a quantum dot populated with an odd number of electrons [23] unusable in flux qubit circuits. The π-junctions can be realised in such systems through a weak Kondo effect [24]. In this case the condensate electrons flip their spin when tunneling through the dot. This process changes the sign of the condensate wave function, *i.e.* introduces a phase shift of π.

Historically, the first system where a Josephson π-junction was realised was a mesoscopic superconductor - normal metal - superconductor (SNS) junction with nonequilibrium current injection [25]. The junction schematics is shown in Fig. 1.6. The control voltage across the gold channel induces a position-dependent electron distribution that affects Josephson coupling energy. Applying some control voltage we can make the Josephson coupling energy to be negative, consequently turning the SNS junction into a π-junction. The SNS junction stays in the π-state, as long as there is the control voltage applied. Therefore, this type of π-junctions can hardly be used in qubit ciruits. Due to stronger coupling to the environment, the coherence times here might be severely limited.

Long before the first physical realisation of a π-junction it was suggested that using superconductor - ferromagnet - superconductor (SFS) sandwiches makes it possible to create a π-junction [26]. Here the π-state can be fixed during the fabrication stage and no control signal is needed to keep the junction in this state. Inversion of the condensate wavefunction sign over the SFS junction occurs due to exchange-field-induced oscillations of superconducting order parameter (the condensate wavefunction) in the ferromagnetic layer. The induced superconductivity in F-layer is spatially inhomogeneous and the superconducting order parameter contains nodes where the phase changes by

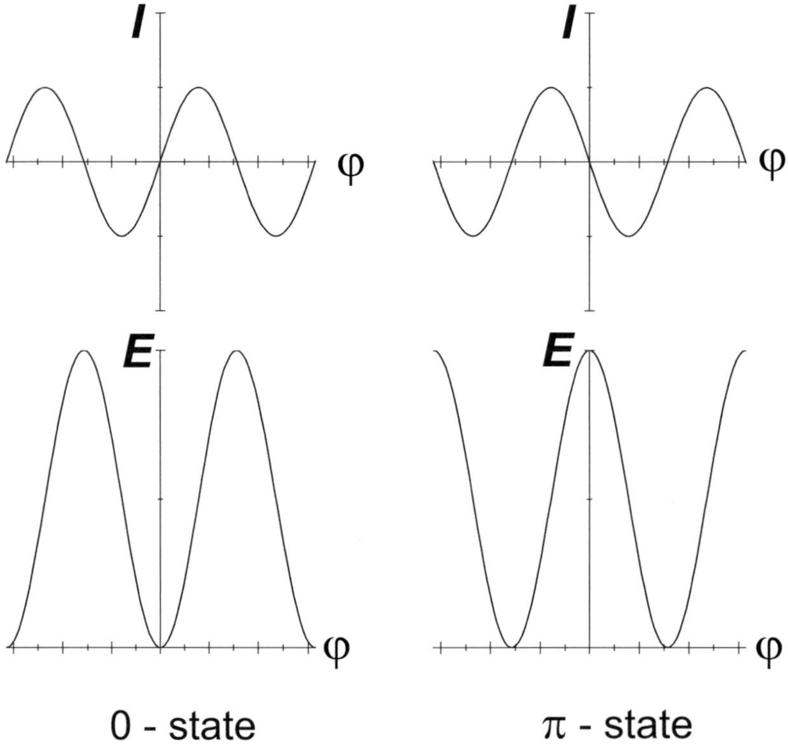

Figure 1.5: The current-phase relations and the energy vs. junction phase dependences of Josephson junctions in conventional (0-) and π-states.

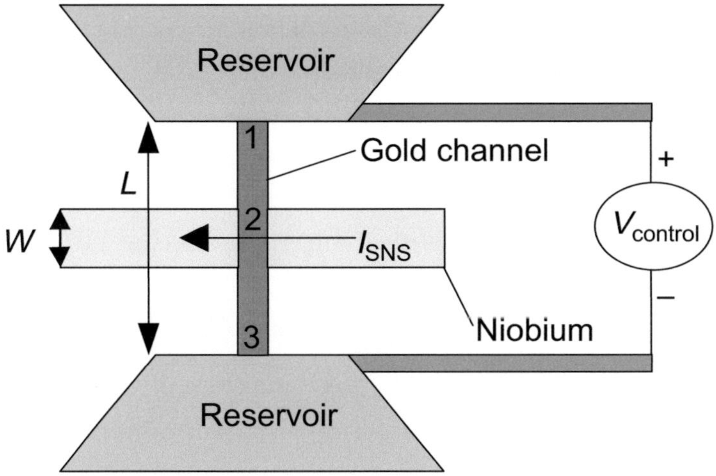

Figure 1.6: A mesoscopic SNS π-junction [25]. A gold channel between two electron reservoirs is connected to two niobium superconducting leads. The control voltage across the channel induces a position-dependent electron distribution that affects Josephson coupling energy. The current through the Josephson junction is indicated by I_{SNS}.

π. Whether the SFS junction will be in 0 or π-state, depends on the thickness of the ferromagnetic interlayer. The advantage of this type of π-junctions is that the phase drop of π appears without any manipulations, if the ferromagnetic thickness is chosen properly. The remaining uncertain issue here is that the ferromagnetic layer might be a source of extra decoherence for a qubit circuit.

Flux qubit experiments with the phase bias provided by flux trap-loops are described in Chapters 2 and 3. Phase shifters based on SFS π-junctions will be discussed in detail in Chapter 4.

1.8 Thesis overview

The original goal of this research was to realise a tunable gap flux qubit biased by an SFS π-junction inserted in the loop. The SFS junctions are fabricated out of niobium, hence we started to explore niobium based flux qubits. Meanwhile, we developed the design of a "silent" tunable flux qubit.

A description of this system and the experimental results are presented in Chapter 2. Due to some fabrication related problems, our attempts to realise all-niobium flux qubit failed. Hence, we tried out aluminum as the qubit material to check if a tunable flux qubit can have relatively long coherence times. This experiment is described in Chapter 3. Since it was doubted that SFS π-junctions can become useful in superconducting circuits designed for quantum computing applications, we endeavoured to check their impact on the coherence properties of the qubits. The details of this experiment can be found in Chapter 4.

References

[1] Einstein, A. Über einen die Erzeugung und Verwandlung des Lichtes betreffenden heuristischen Gesichtspunkt. *Annalen der Physik* **17,** 132-148 (1905).

[2] Planck, M. Ueber das Gesetz der Energieverteilung im Normalspectrum. *Annalen der Physik* **4,** 553-563 (1901).

[3] Einstein, A., Podolsky, B. & Rosen, N. Can quantum-mechanical description of physical reality be considered complete? *Phys. Rev.* **47,** 777-780 (1935).

[4] Aspect, A., Grangier, P. & Roger, G. Experimental realization of Einstein-Podolsky-Rosen-Bohm Gedankenexperiment: a new violation of Bell's inequalities. *Phys. Rev. Lett.* **49,** 91-94 (1982).

[5] Deutsch, D. Quantum theory, the Church Turing principle, and the universal quantum computer. *Proc. Roy. Soc. Lond. A* **400,** 97-117 (1985).

[6] Turing, A. M. On computable numbers, with an application to the entscheidungsproblem. *Proc. London Math. Soc.* **42,** 230-265 (1937).

[7] Shor, P. Algorithms for quantum computation: Discrete logarithms and factoring. *Proceedings of the 35th Annual IEEE Symposium on Foundations of Computer Science,* 124-134 (1994).

[8] DiVincenzo, D. P. The physical implementation of quantum computation. *Fortschritte der Physik* **48,** 771-783 (2000).

[9] Doll, R. & Näbauer, M. Experimental proof of magnetic flux quantization in a superconducting ring. *Phys. Rev. Lett.* **7,** 51-52 (1961).

[10] Josephson, B. D. Possible new effects in superconductive tunnelling. *Physics Letters* **1,** 251-253 (1962).

[11] Nakamura, Y., Pashkin, Y. A. & Tsai, J. S. Coherent control of macroscopic quantum states in a single-Cooper-pair box. *Nature* **398,** 786Ű788 (1999).

[12] Martinis, J. M., Nam, S., Aumentado, J. & Urbina, C. Rabi oscillations in a large Josephsonjunction qubit. *Phys. Rev. Lett.* **89,** 117901 (2002).

[13] Vion, D. *et al.* Manipulating the quantum state of an electrical circuit. *Science* **296,** 886Ű889 (2002).

[14] Koch, J. *et al.* Charge-insensitive qubit design derived from the Cooper pair box. *Phys. Rev. A* **76,** 042319 (2007).

[15] Chiorescu, I., Nakamura, Y., Harmans, C. J. P. M. & Mooij, J. E. Coherent quantum dynamics of a superconducting flux qubit. *Science* **299,** 1869-1871 (2003).

[16] Clarke, J. & Wilhelm, F. K. Superconducting quantum bits. *Nature* **453,** 1031-1042 (2008).

[17] Orlando, T. P. *et al.* Superconducting persistent-current qubit. *Phys. Rev. B* **60,** 15398-15413 (1999).

[18] Ithier, G. *et al.* Decoherence in a superconducting quantum bit circuit. *Phys. Rev. B* **72,** 134519 (2005).

[19] Golubović, D. S., Pogosov, W. V., Morelle, M. & Moshchalkov, V. V. Magnetic phase shifter for superconducting qubits. *Phys. Rev. Lett.* **92,** 177904 (2004).

[20] Majer, J. B., Butcher, J. R. & Mooij, J. E. Simple phase bias for superconducting circuits. *Appl. Phys. Lett.* **80,** 3638-3640 (2002).

[21] Van Harlingen, D. J. Phase-sensitive tests of the symmetry of the pairing state in the high-temperature superconductors—Evidence for $d_{x^2-y^2}$ symmetry. *Rev. Mod. Phys.* **67,** 515-535 (1995).

[22] Bulaevskii, L. N., Kuzii, V. V. & Sobyanin, A. A. Superconducting system with weak coupling to the current in the ground state. *JETP Lett.* **25,** 290-294 (1977).

[23] Cleuziou, J.-P., Wernsdorfer, W., Bouchiat, V., Ondarcuhu, T. & Monthioux, M. Carbon nanotube superconducting quantum interference device. *Nature Nanotech.* **1,** 53-59 (2006).

[24] Glazman, L. I. & Matveev, K. A. Resonant Josephson current through Kondo impurities in a tunnel barrier. *JETP Lett.* **49,** 659-662 (1989).

[25] Baselmans, J. J. A., Morpurgo, A. F., van Wees, B. & Klapwijk, T. M. Reversing the direction of supercurrent in a controllable Josephson junction. *Nature* **397,** 43-45 (1999).

[26] Buzdin, A. I., Bulaevskij, L. N. & Panyukov, S. V. Critical-current oscillations as a function of the exchange field and thickness of the ferromagnetic metal (F) in an S-F-S Josephson junction. *JETP Lett.* **35,** 178-180 (1982).

Chapter 2

Measurements of two-cell flux qubits

The flux qubits use junctions enclosed in superconducting loops as quantum coherent systems. *Single-cell* flux qubits have been previously studied at MIT and Delft [1]. These devices have already shown quantum superposition of macroscopic persistent current states and quantum-coherent oscillations in direct time-resolved measurements. We are focusing on *two-cell* flux qubits consisting of four Josephson tunnel junctions. These devices have an advantage of having a control over the barrier hight between two potential wells.

2.1 Introduction

The standard Delft [1] qubit based on the single-loop 3-junction SQUID has a double well potential at half frustration. It is not possible to change the splitting between the two lowest energy levels of the qubit during an experiment without breaking the symmetry of the potential, since the barrier height at half frustration is fixed by fabrication. Outside the symmetry point the flux qubit is linearly coupled to flux noise that results in short dephasing time. In the original persistent-current qubit [2] shown in Fig. 2.1(a) the height of the barrier at the symmetry point can be tuned during an experiment, but one has to change two control fluxes simultaneously to keep the symmetry of the potential. This will also require a stronger coupled control line which can in turn reduce qubit coherence. The full tunability of this circuit is reached by replacing the smaller qubit junction by a dc-SQUID, thus its Josephson energy and consequently the minimal splitting between the two lowest energy levels of the qubit (the qubit gap) can be tuned by varying an externally applied flux Φ_α.

At the symmetry point the flux qubit dephasing time is maximal. Therefore, it would be favourable to develop a qubit design where the splitting between the two lowest energy levels of the qubit can be changed *in situ* during an experiment by changing a current flowing in one control line without affecting the symmetry of the potential. That is we are looking for a system with an orthogonal control over the Hamiltonian

$$H = -\frac{\varepsilon}{2}\sigma_z - \frac{\Delta}{2}\sigma_x \, , \tag{2.1}$$

Such qubit will allow us to realise a switchable coupling between the qubit and a cavity keeping the qubit coherence optimal (please see Chapter~\ref{chap:osc} for details on this experiment).

In Fig. 2.1(b,c), we schematically present two topologically modified versions of the tunable gap flux qubit presented in Fig. 2.1(a). Both structures features an orthogonal control over the Hamiltonian (2.1), because of the internal symmetry of the circuits.

For our initial experiments we have chosen the version shown in Fig. 2.1(c), which we call a two-cell flux qubit, since it is possible to manipulate

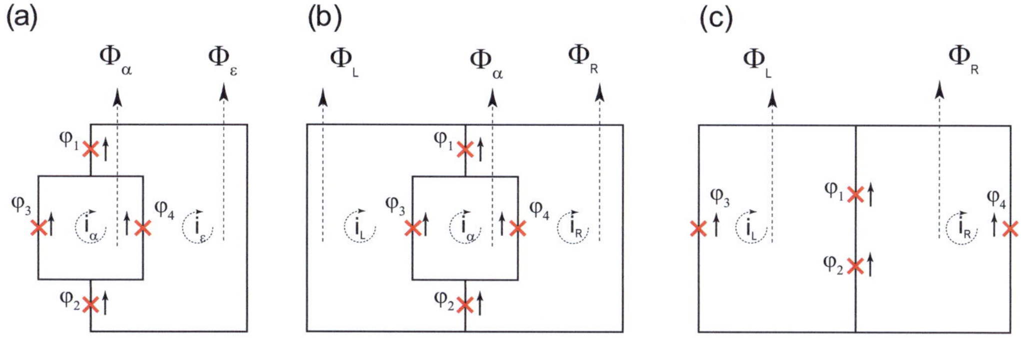

Figure 2.1: Tunable gap flux qubit designs. (a)Original proposal by J. E. Mooij *et al.* [2] (b) Gradiometer tunable gap flux qubit. (c) Two-cell flux qubit

it without an external superconducting coil.

We came to the idea of a two-cell flux qubit through a collaboration with Dr. Stanford Yukon, when we were investigating the triangular long Josephson junction prism discussed in Refs. [3, 4].

2.2 Qubit potential

A two-cell flux qubit shown in Fig. 2.1(c) contains four Josephson junctions of two different sizes. Two outer Josephson junctions numbered #3 and #4 have smaller area than two inner ones numbered #1 and #2. We denote by ρ the ratio between their critical currents and thus their Josephson energies:

$$E_J^{(3,4)} = \rho\, E_J^{(1,2)}, \tag{2.2}$$

The potential energy U of the qubit is periodic as a function of the average phase $\chi = 0.5(\varphi_1 + \varphi_2)$ on junctions #1 and #2. Assuming that we are working in the regime of a well defined phase $|\varphi_1 - \varphi_2| \ll 1$ and the potential energy can be written as:

$$U = -2\cos\chi + r\cos(2\chi - 2\pi\frac{\phi_a}{\Phi_0})\,. \tag{2.3}$$

$$r = -2\rho\cos(2\pi\frac{\phi_s}{\Phi_0}), \tag{2.4}$$

while

$$\phi_s = \frac{\phi_L + \phi_R}{2}$$

and

$$\phi_a = \frac{\phi_L - \phi_R}{2}$$

are the symmetric and anti-symmetric parts of the flux in the two cells, respectively. Each period of $U(\chi)$ has two minima. The height of the barrier between these minima is defined by r (see Fig. 2.2) that can be changed *in situ* by changing the symmetric component of the flux applied to the qubit loop ϕ_s. For $\rho = r/2$ we can operate this qubit at half frustration in each cell, *i.e.* an integer number of flux quanta penetrating the outer qubit loop.

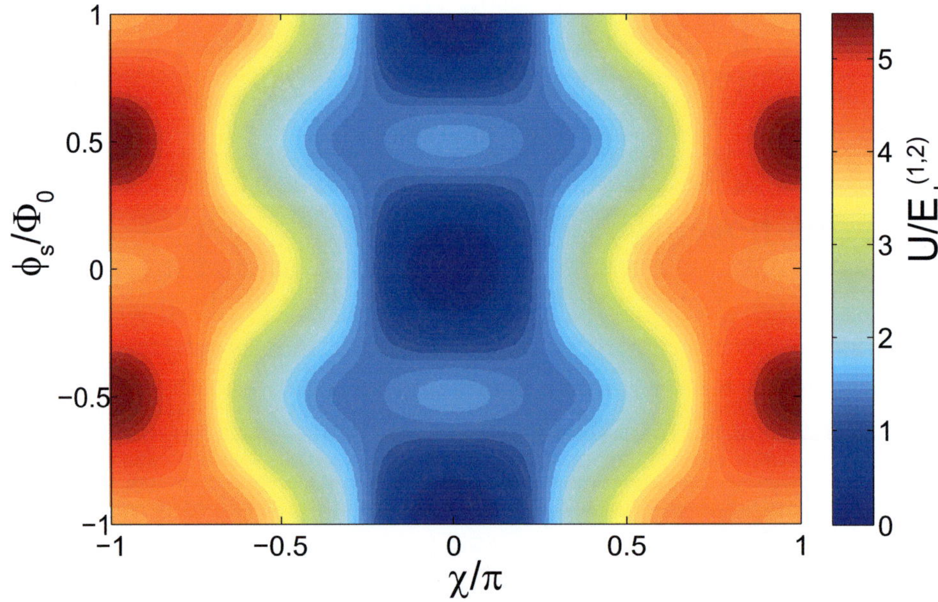

Figure 2.2: The two-cell flux qubit potential $U(\chi, \phi_s)$.

2.3 "Silent" two-cell qubit

We developed a modified scheme for two-cell flux qubits which we call a "silent" qubit. Its main idea can be seen from Fig. 2.3. With a single flux quantum Φ_0 frozen in the inner loop the qubit has the advantage of only smaller range of magnetic fields needed for the symmetric flux control. The trapped flux serves as π-shifter of the phase [5] and contributes equally to the phase change in both loops. It corresponds to $\phi_a/\Phi_0 = 1/2$ that can be considered as simply a change of the sign of r in the potential (2.3). Thus one does not need to apply large external flux and the qubit is operated close to zero symmetric field.

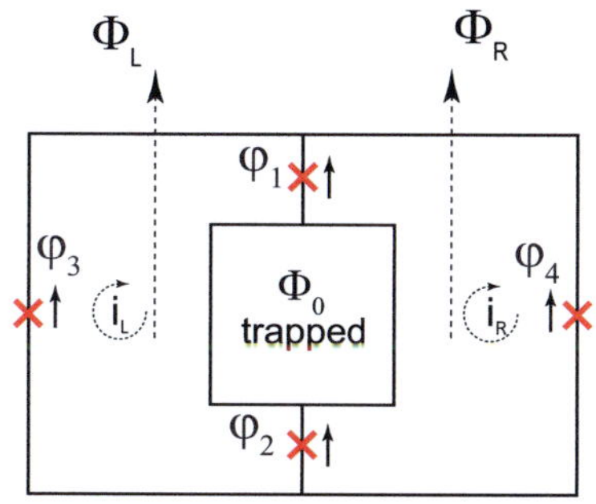

Figure 2.3: A "silent" two-cell flux qubit

The potential (2.3) has two minima associated with the circulating currents in the outer loops, as shown in Fig. 2.4. Numerical calculations show that at the working point of $r = 2\rho \approx 0.8$, the current through the outer qubit junctions nearly reaches the critical current value (*i.e.* is maximal). When qubit switches from one persistent current state to the other (*i.e.* from state (a) to state (b)in Fig. 2.4), the current through the central line changes by approximately $4I_c^{(3,4)} = 2rI_c^{(1,2)}$, thus the state of the qubit can be distinguished by a coupled dc-SQUID. The dc-SQUID should be coupled asymmetrically to the qubit cells, because of the internal symmetry of the qubit.

In quantum regime, at the qubit symmetry point the mean current through every Josephson junction is zero due to quantum superposition of two current states. Taking into account that we work near zero external magnetic field, it means that at the operating point the "silent" qubit is decoupled in the linear order from the fluctuations caused by the flux control and readout lines.

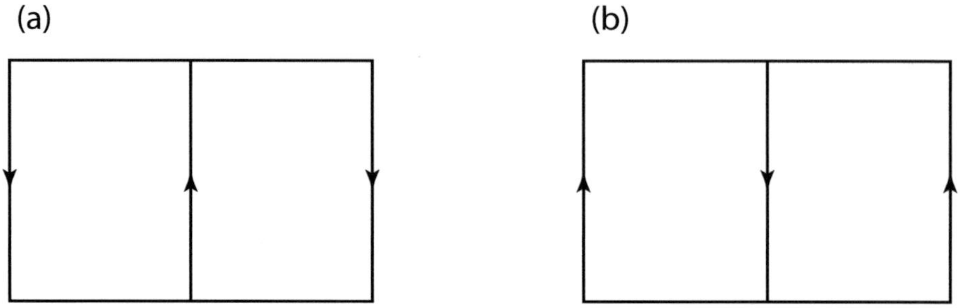

Figure 2.4: Two states of the simplified version of the two-cell flux qubit shown in Fig. 2.1(c). The arrows show the flow directions of the persistent currents.

Generally speaking, the flux trap-loop in Fig. 2.3 can be replaced by any structure that provides a phase difference of π between its terminals, *e.g.* an SFS π-junction [6, 7]. In distinction from phase-shifting loops with frozen magnetic flux [5], the SFS circuits do not require trapping a well-defined integer number flux quanta in their superconducting loops. these alternative phase shifters based on SFS π-junctions will be discussed in detail in Chapter 4.

However, the qubit design version shown in Fig.2.3 has also some disadvantages. In this scheme a galvanic coupling between the qubit and the readout SQUID is not possible. To achieve a required readout contrast, one has to make qubit loop size larger. This also increases coupling to the environment and consequently decreases qubit coherence. Superconducting islands between junction #1 and junctions #3, #4 are large, thus the qubit might be more sensitive to charge noise, because of stronger capacitive coupling to the control circuitry. Moreover, a small trap-loop can trap a single flux quantum only when it is made of a superconductor with a high depairing current, *e.g.* niobium, but it is difficult to fabricate submicron-scale niobium tunnel junctions: the technological process is very complicated and has some limitations.

Up to now the easiest and the most reliable technique to fabricate submicron-scale Josephson tunnel junction has been the Dolan process [8] also known as two angle shadow evaporation. This process allows for fabricating of small Josephson junctions and simple circuits in a single lithography step. Usually this process is used to produce $Al - AlO_x - Al$ tunnel junctions, because of low melting point of aluminium and reasonable quality of the AlO_x tunnel barriers.

Since a trap-loop made of aluminium should be relatively large, we cannot use the scheme described above. An outer trap-loop (see Fig. 2.1(b)) is a better choice in this case. It allows us to keep the qubit loop inductance small, which in turn decreases coupling to the environment. The qubit junctions are placed close to each other, which should decrease sensitivity to charge noise. This system features a gradiometer design, fluctuations of the uniform magnetic field affect only the height of the barrier due to the circuit geometry. The corresponding fluctuations of the barrier height will be small, as long as the size of the inner loop is small. An experiment with an aluminium gradiometer qubit will be described in Chapter 3.

2.4 Initial design

We designed a readout circuit to measure the “silent” qubit described above. This circuit contains an asymmetric three-junction dc-SQUID inductively coupled to the qubit. The SQUID has non-zero flux sensitivity at zero external magnetic field. The qubit is placed inside the readout SQUID loop, which is coupled asymmetrically to the two cells of the qubit. The flux through the SQUID loop changes when the qubit switches between two persistent current states.

We submitted our design to MIT Lincoln Laboratory foundry. The submitted layout contains circuits discussed in the previous section. Sizes of junctions were optimised for the critical current density of 500 A/cm^2.

The targeted critical current I_{Cq} of inner qubit junctions is about 0.35 μA and capacitance $C_q \approx 5$ fF. The outer qubit junctions are designed 2.33 times smaller than the inner ones. The calculated splitting between the two of the lowest levels of the qubit is around $\nu_{01} \approx 2.5$ GHz at the working point $r = 0.8$ (see Eq. (2.4) and the calculated transition frequency between the first and the second excited states is $\nu_{12} \approx 27.5$ GHz.

A fragment of the circuit layout is presented in Fig. 2.5. In this circuit we intended to use biasing resistors on chip (shown in green). We have placed on chip three different types of control lines: (i) an antisymmetric line to change antisymmetric part of the flux through the qubit cells, *i.e.* the symmetry of the qubit potential (in Fig. 2.5 it is situated below the qubit), (ii) a symmetric line to control the working point r, *i.e.* the hight of the potential barrier (above the qubit in Fig. 2.5), and (iii) two lines for biasing the readout SQUID (to the left of the qubit in Fig. 2.5).

Every control line is weekly coupled to the qubit in order to minimise coupling of the noise in the control lines to the qubit.

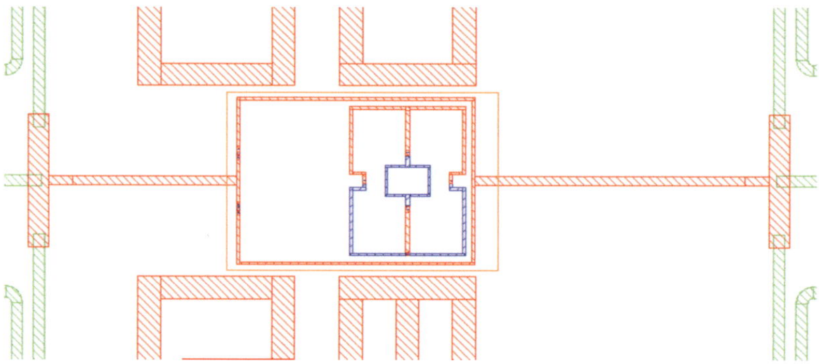

Figure 2.5: A fragment of the circuit layout.

We have designed three circuits with different coupling strength between the readout SQUID and the qubit: "Strong","Medium" and "Weak". Qubit parameters were the same for every circuit. The self inductance of every loop of the two-cell flux qubit is $L_{QB} \approx 155\,\text{pH}$. Mutual inductances between a cell of the qubit and the control lines are:

- for the antisymmetric control line $M_{QB}^{(AS)} \approx 0.55\,\text{pH}$;

- for the symmetric control line $M_{QB}^{(S)} \approx 2.8\,\text{pH}$.

Every readout SQUID consists of three Josephson junctions. There is a junction with the critical current $I_C = I_{Cq}$ in one branch and two junctions in series with twice larger critical current in the other branch. Thus the inductances of the SQUID branches are equal, that should decouple the qubit from the room temperature electronics at zero bias of the readout SQUID. Parameters of the readout SQUID for every coupling strength are summarised in the following table.

	"Strong"	"Medium"	"Weak"
M	22.9 pH	17.3 pH	12.8 pH
$\Delta\Phi_{max}$	$3.1\,\mathrm{m}\Phi_0$	$2.4\,\mathrm{m}\Phi_0$	$1.8\,\mathrm{m}\Phi_0$
L_{SQUID}	355 pH	360 pH	364 pH
$M^{(AS)}_{SQUID}$	0.89 pH	0.85 pH	0.80 pH
$M^{(S)}_{SQUID}$	11.2 pH	11.3 pH	11.4 pH

The parameters are:

- mutual inductance between the qubit and the readout SQUID M;
- maximum change of the flux penetrating the readout SQUID loop induced by the qubit switching $\Delta\Phi_{max} = 2M\rho I_{Cq}$;
- self inductance L_{SQUID};
- mutual inductance between the readout SQUID and the antisymmetric control line $M^{(AS)}_{SQUID}$;
- mutual inductance between the readout SQUID and the symmetric control line $M^{(S)}_{SQUID}$.

2.5 Measurements of the samples from the wafers #1 and #2

2.5.1 Samples

We have received two batches of $Nb/AlO_x/Nb$ junction chips fabricated according to our design at MIT Lincoln Laboratory. Every chip had three qubit circuits with different coupling strength between the readout SQUID and the qubit and four single Josephson junctions of the same sizes as junctions in the qubit and the readout SQUID. We use them for testing purposes. Two test junctions have the size equal to the size of the smallest qubit junctions.

Figure 2.6 shows an SEM micrograph of a two-cell flux qubit together with its readout SQUID. Red circles mark regions where Josepson junctions are situated.

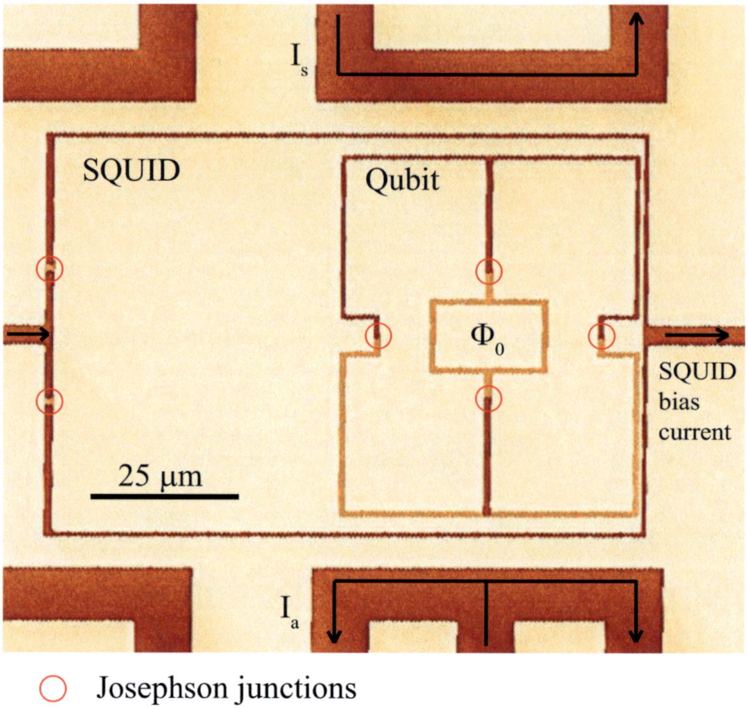

Figure 2.6: An SEM micrograph of a sample fabricated at MIT Lincoln Laboratory.

The fabrication process is characterised by a process bias, which determines a reduction of the electrical junction sizes form the drawn sizes. The average process bias specified by Lincoln Laboratory in the design rules is about 0.25 μm. This value was taken into account when designing the qubits. Larger process biases result in increasing the maximal barrier hight and decreasing of the qubit flux signal. Smaller process biases result in decreasing of the maximal barrier hight, this makes qubit operating difficult. However, there is a variation across the wafer and from wafer to wafer. For the wafer #1 the process bias is to $0.23 \pm 0.03\,\mu$m. The dies with the process bias close to 0.25 μm had smaller yield of junctions. The yield for those chips is seven working junctions out of eleven drawn, the median yield over the wafer is nine working junctions. The yield estimate includes the drawn 0.2 μm and 0.3 μm Josephson junctions. The smallest junction reliably yielded has the drawn size of 0.4 μm, while the smallest qubit junction has the drawn size of 0.45 μm. For the wafer #2 the process bias is $0.21 \pm 0.04\,\mu$m and the median yield is eight working junctions out of nine drawn. The smallest junction has the drawn size of 0.4 μm. The chip B4 is the only die with the process bias equal to 0.25 μm on this wafer. There are several dies with process bias 0.26 μm and 0.27 μm on the wafer, but the expected qubit signals on these chips might be too week, as the critical current density on these dies is around 440 A/cm^2. Thus only few chips from the available batch were expected to be close to the targeted parameters.

The test Josephson junctions showed high-quality current-voltage characteristics, which are presented in Fig. 2.7. By scaling these curves we can calculate sizes of the junctions. The small qubit junctions have area of about 0.03 μm^2, the large qubit junctions are of about 0.08 μm^2 and the large SQUID junctions have the area of about 0.16 μm^2. These values are close to designed values, which were 0.03 μm^2, 0.07 μm^2 and 0.14 μm^2 respectively, and suitable for qubit operation. One can see that the spread between critical currents of two smallest qubit junctions is negligible.

Figure 2.8 shows measured current-voltage characteristics of readout SQUIDs on the chip C4 from the wafer #1 and on the chip B4 from the wafer #2. Despite the high quality of the test junctions, the critical current of the readout SQUID is much below the expected value. There is no peculiarity around 3 mV, which should correspond to $2\Delta/e$, where Δ is the superconducting energy gap of niobium. We find a step around 1 mV for the wafer #1 and 0.4 mV for the wafer #2, which can most probably be due to difference of energy gaps in top and bottom niobium electrodes. We measured six chips from both wafers (we tested chips with the process bias about 0.25 μm), and all 15 measured readout SQUIDs had the current-voltage characteristics similar to ones presented in Fig. 2.8. A couple of SQUIDs showed clear indications of shorts, but they might have been destroyed by an electro-static discharge during mounting procedure. This behaviour of the readout SQUIDs was later confirmed by tests of 42 SQUIDs done by Dr. William Oliver at

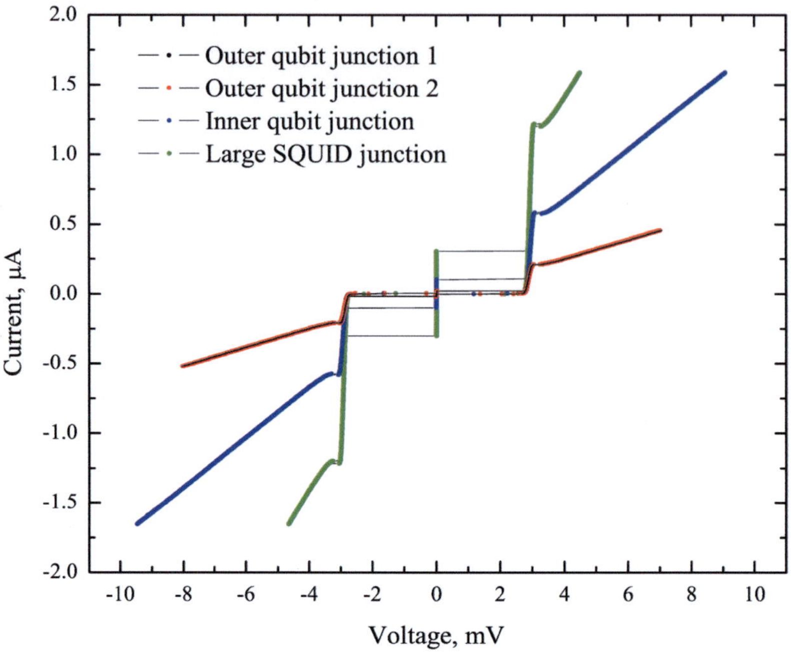

Figure 2.7: Current-voltage characteristics of the test junctions on chip C4 from the wafer #1 at 300 mK.

MIT Lincoln Laboratory. He assumed that these defects are caused by the sizes of the junctions in the readout SQUIDs and the topology of electrodes which were creating a kind of antenna for plasma throughout inductively coupled plasma etching steps of the fabrication process. Dr. William Oliver also suggested that we should enlarge junctions in the readout SQUID. This was done in May 2008 when we submitted the revised circuit design to Lincoln Laboratory foundry.

In spite of the abnormal behaviour of the readout SQUIDs we were able to observe a modulation of the SQUID critical current with applied magnetic flux. This allows us to read the qubit state out around zero symmetric control flux.

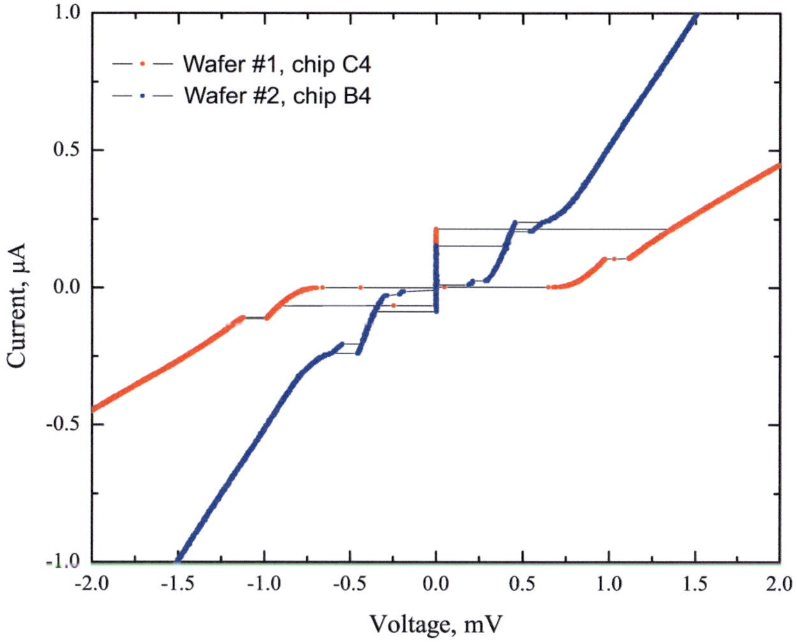

Figure 2.8: Current-voltage characteristics of the readout SQUIDs at 300 mK.

2.5.2 Characterisation procedure

First of all we characterised the readout SQUID. For this purpose we measure a dependence of the SQUID critical current (I_c) on the symmetric control flux (I_s). This dependence allows us to estimate the readout SQUID sensitivity around zero applied flux. If the sensitivity is too low we can try to increase it by flux biasing the readout SQUID using one of the SQUID biasing lines.

To bias a two-cell flux qubit to the optimal working point $r = 2\rho$ (assuming we have trapped an even number of flux quanta in the passive π-shifter loop) we should apply a flux quantum to the qubit loop (*i.e.* a half flux quantum per qubit cell). Taking into account that the mutual inductance between the symmetric control line and the qubit loop is almost twice smaller than the mutual inductance between the symmetric control line and the readout SQUID one can readily see that the optimal working point of the qubit corresponds to two flux quanta penetrating the readout SQUID loop. Thus subtracting a dependence of the SQUID critical current on the antisymmetric control flux (I_a) at zero symmetric control flux from one at the optimal working point, we should see an “S”-curve which corresponds to switching of the qubit between two persistent current states. If we cannot see the “S”-curve it means that the qubit has a defect.

If we have trapped an odd number of flux quanta in the passive π-shifter loop while cooling down, the qubit will be biased to the optimal working point at zero symmetric control flux and the qubit potential will have two minima. In this case a flux quantum applied to the qubit loop will turn the qubit potential into single-well one and the difference of the dependences of the SQUID critical current on the antisymmetric control flux at these symmetric control fluxes should show us qubit switching between two persistent current states.

2.5.3 Measurement technique

To flux bias a qubit we use custom built battery powered current sources. These sources are controlled by a voltage in the range of $\pm$ 10 V and generate currents in a switchable range from 10^{-2} to 10^{-7} A/V. The dc-voltages corresponding to the flux levels are delivered by a set of 16-bit precision digital-to-analog converters. These are equipped with a serial interface that allows cascading of multiple converters in a daisy-chain, such that only one signal input is necessary. To avoid the presence of digital noise in the electronics during qubit operation, the DAC clock signal is generated by a custom-built logic circuit only during reprogramming the voltage levels. The digital data are encoded in a pulse-width modulated signal, which is created by a National Instruments (NI) output interface card in the controlling computer [9].

The current source for biasing the readout SQUID is controlled by a saw-tooth generator, which produces a linearly increasing voltage once a trigger signal is sent to its start input. The voltage across the SQUID is amplified by a factor of 1000 using an instrumentation amplifier at room temperature, whose output is fed into a trigger calibrated to generate a TTL pulse whenever the SQUID switched to its voltage state. This pulse is applied to the stop input of the ramp generator to switch the SQUID within approximately 200 μs back to the superconducting state, thus avoiding excess heating of the sample chip. The time Δt between the pulses starting the current ramp and the voltage trigger is measured by a precision time interval counter. With the knowledge of the current ramp rate dI/dt, the switching current $I_{sw} = \frac{dI}{dt}\Delta t$ is obtained with high resolution. This sequence is repeated at typical rates of 200 to 400 Hz, which are limited by the time necessary to do the current-ramp measurement and subsequent idle time for cool-down, in total of about 2 ms. One data point for a given parameter set is usually averaged over 1000 to 5000 repetitions.

The described technique was used to measure every qubit circuit except for circuits on the chip C4 from the wafer #1. Due to a lack of high precision electronics throughout experiments with this die we used the National Instruments output interface card to supply dc-voltages that define flux levels. This card was also controlling a current source used for current biasing of the readout SQUID. The switching current was measured by an analog to digital converter of the NI interface card.

2.5.4 Results

Since the resolution of our detectors was limited due to fabrication related problems, we only measured structures with "strong" coupling between the redout SQUID and the qubit, because of a larger signal induced by the qubit switching and thus larger probability to detect it. We chose dies with a high yield of test junctions and the process bias close to 0.25 μm. The data below is reported to illustrate the performed work. Unfortunately no successful results were obtained besides a qubit switching between two classical states.

Wafer #1, die C4

The die C4 was used to check quality of the test junctions and readout SQUIDs. The chip was measured in a Helium-3 cryostat at a temperature of about 300 mK. As was mentioned above the readout SQUIDs showed abnormal behaviour, which was limiting resolution of our readout circuits. We also tried to characterised qubits, but the technique used to measure the SQUID critical current had low resolution and thus the results are not conclusive.

Wafer #1, die C3

The chip #C3 was measured in a dilution refrigerator at the base temperature of about 50 mK. The Fig. 2.9 shows the dependence of the readout SQUID critical current on the symmetric control field. We saw no difference between the dependence of the SQUID critical current on the antisymmetric control flux at zero applied symmetric flux and the dependence with one flux quantum penetrating the qubit loop. That indicates that the qubit has a defect. The measured curves are shown in Fig. 2.10.

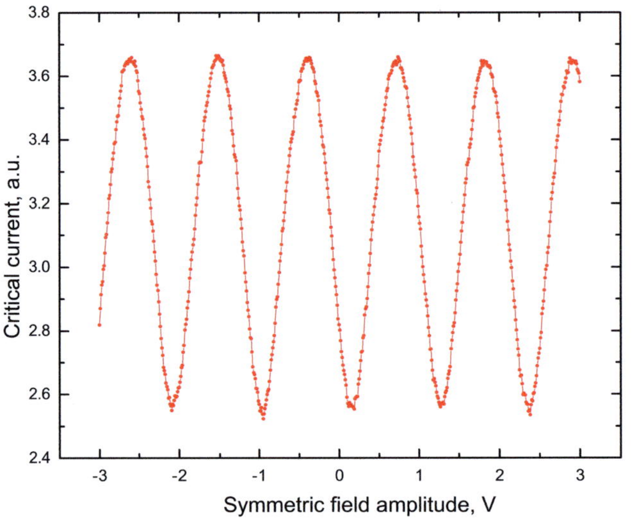

Figure 2.9: Dependence of the critical current of the readout SQUID on the chip C3 from the wafer #1 on symmetric control flux.

Wafer #2, die D2

The measurements of the die D2 were also done in a dilution refrigerator at its base temperature of 50 mK. The I_c *vs.* I_s dependence is shown in Fig. 2.11. The Fig. 2.12 shows I_c *vs.* I_a dependences for different values of applied symmetric flux. The curves show no qubit signature.

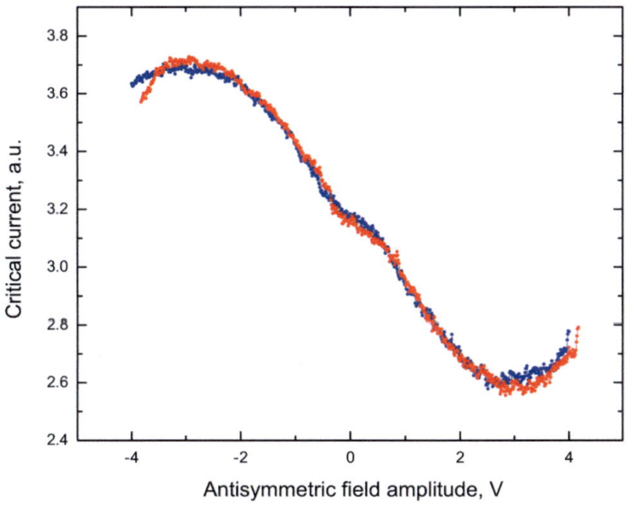

Figure 2.10: Dependence of the critical current of the readout SQUID on the chip C3 from the wafer #1 on antisymmetric control flux.

Wafer #2, die D4

The die D4 was measured in our helium-3 cryostat at the temperature of about 300 mK. The Figs. 2.13 and 2.14 show $I_c\, vs.\, I_s$ and $I_c\, vs.\, I_a$ depen-

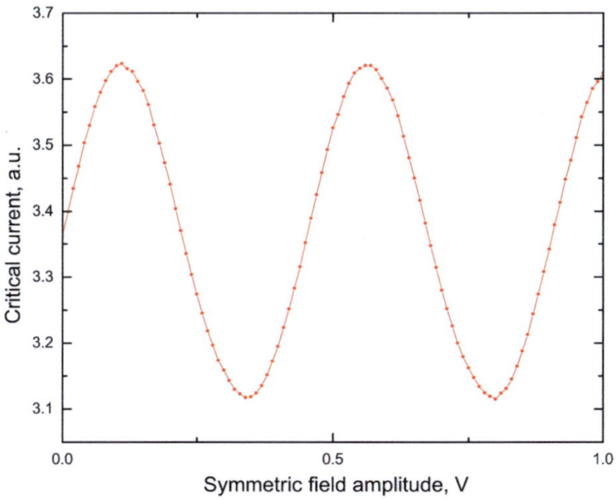

Figure 2.11: Dependence of the critical current of the readout SQUID on the chip D2 from the wafer #2 on symmetric control flux.

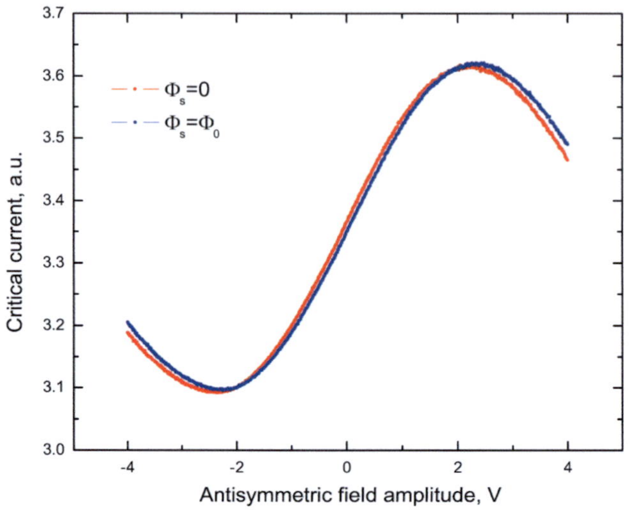

Figure 2.12: Dependence of the critical current of the readout SQUID on the chip D2 from the wafer #2 on antisymmetric control flux.

dences respectively. No feature which could be associated with the qubit switching was observed.

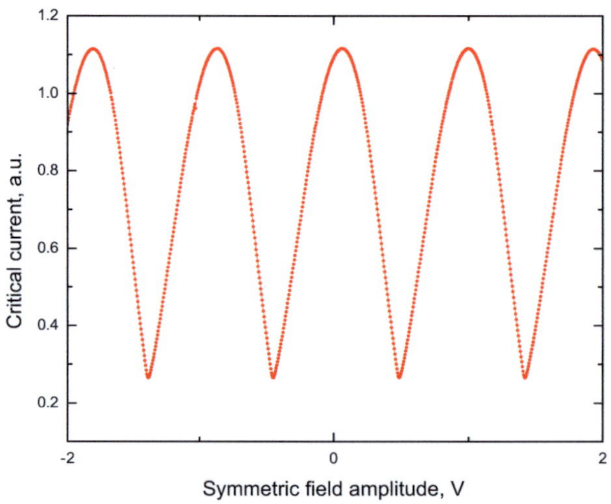

Figure 2.13: Dependence of the critical current of the readout SQUID on the chip D4 from the wafer #2 on symmetric control flux.

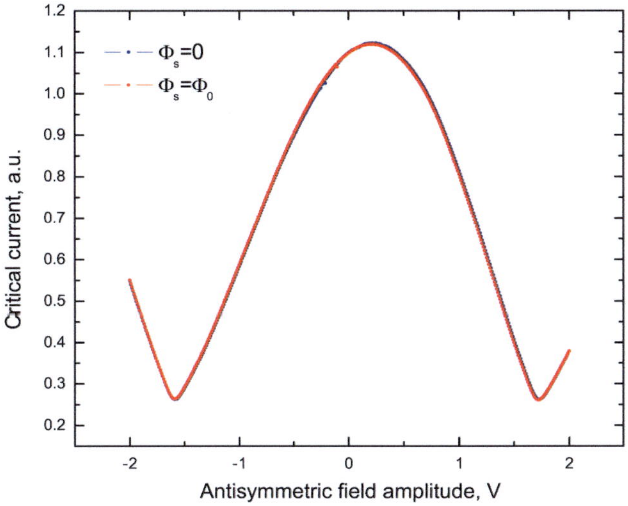

Figure 2.14: Dependence of the critical current of the readout SQUID on the chip D4 from the wafer #2 on antisymmetric control flux.

Wafer #2, die B4

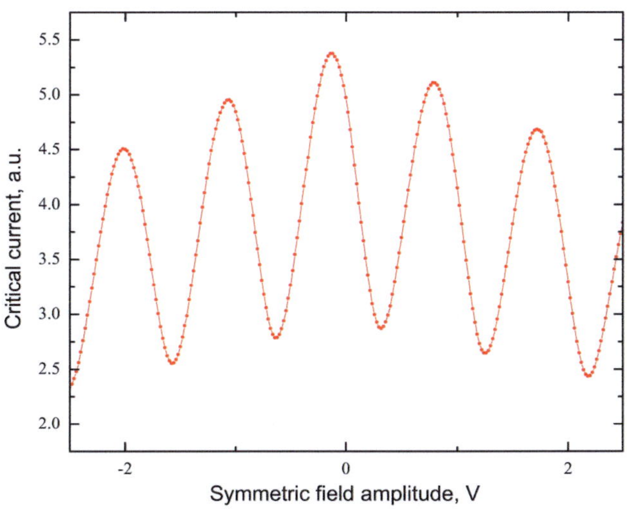

Figure 2.15: Dependence of the critical current of the readout SQUID on the chip B4 from the wafer #2 on symmetric control flux.

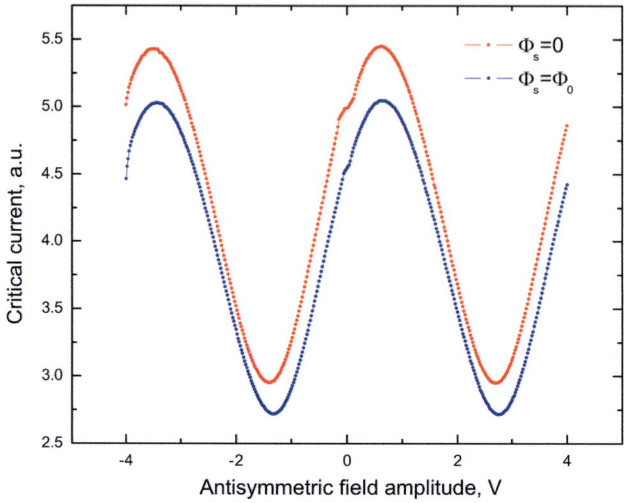

Figure 2.16: Dependence of the critical current of the readout SQUID on the chip B4 from the wafer #2 on antisymmetric control flux.

The dependence of the SQUID critical current on symmetric control flux induced by control line I_s for the readout SQUID on the chip B4 is presented

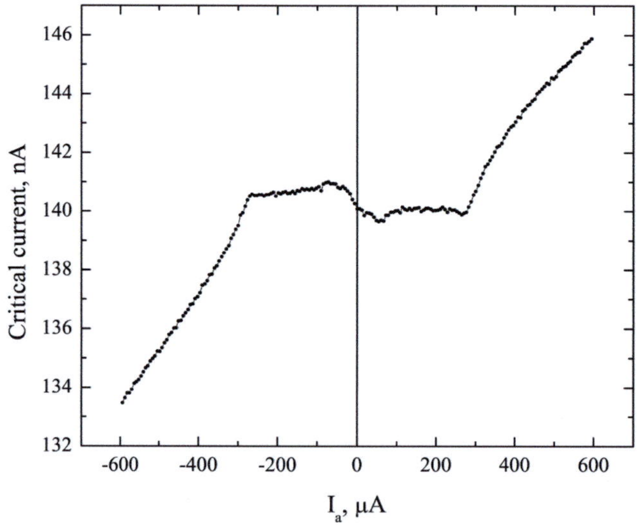

Figure 2.17: Dependence of the critical current of the readout SQUID on the chip B4 from the wafer #2 on antisymmetric control flux.

in Fig. 2.15. The measurements of this die were done at the temperature of about 300 mK. The figure shows that the critical current of the SQUID is decreasing with the applied symmetric flux. It indicates that we are heating up our circuit by a current flowing through the symmetric control line. Most probably this happens due to a large contact resistance on a contact pad of the line. The $I_c\, vs.\, I_a$ dependences are presented in Fig. 2.16. Since the dependences were taken at different effective temperatures, we cannot compare them directly. However, one can readily see a step-like feature around zero of antisymmetric control flux. This feature is present on both curves.

Figure 2.17 presents the dependence of the readout SQUID critical current on antisymmetric control flux at zero applied symmetric flux measured with high resolution around zero antisymmetric control flux. The dependence is non-monotonic around zero flux. At zero antisymmetric flux, we expect a switching of the qubit from one classical state to another, which changes the magnetic flux penetrating the readout SQUID loop. However, the measured curve does not look like an abrupt switching, as we see a slow change of the flux instead.

We believe that the zero-field response of the circuit is due to an odd number of flux quanta trapped in the passive π-shifter loop while cooling down. We made three attempts to trap flux inside the passive π-shifter loop. Making further tries turned impossible due to rapid degradation of the readout circuit. We saw a shift of the dependence of the SQUID critical current on the control current I_s after every attempt, which means that we trapped a different number of quanta However, every time we trapped an odd number of quanta, as we saw a switching at zero applied symmetric flux.

This feature cannot be explained by a two-cell flux qubit switching, since it does not depend on the value of the applied symmetric control flux.

2.5.5 Discussion

Qubits from four dies from both wafers were characterised using high-resolution current-ramp technique. Every qubit circuit was insensitive to the symmetric control flux, this fact can be explained by the assumption that the outer qubit junctions were damaged (short or open) even on the dies with high yield of test junctions.

2.6 Revised design

Taking into account suggestions of Dr. William Oliver we redesigned readout SQUIDs to avoid damage of the SQUID junctions by plasma etching during fabrication. In the new design the critical currents of junctions used in the readout SQUIDs are 1.4 μA and 2.8 μA. Josephson junctions in a few readout SQUIDs were shunted by capacitors to minimise quantum fluctuations.

All the self and mutual inductances were kept unchanged. Thus the $\beta_L = \frac{2\pi L I_C}{\Phi_0}$ parameter for the readout SQUID is around 1.5 for the new design, this might decrease the sensitivity of the readout circuit.

The revised design was submitted to the Lincoln Laboratory foundry in May 2008. The submitted design contained structures optimised for the critical current densities of 500 A/cm^2 and 1 kA/cm^2.

2.7 Measurements of the samples from the wafer #3

2.7.1 Samples

We have received seven dies from the wafer #3, that was fabricated according to the revised design. Unfortunately we have not got any test data from MIT Lincoln Laboratory. Every chip contains two sets of structures designed for the critical current densities of 500 A/cm^2 and 1kA/cm^2. Every set consists of seven qubit circuits with three different coupling strengthes between the readout SQUID and the qubit and four single Josephson junctions. In every set Josephson junctions in four readout SQUIDs were shunted by capacitors.

A readout SQUID with unshunted Josephson junctions showed a high-quality current-voltage characteristic, however, a readout SQUID with shunted Josephson junctions fabricated on the same chip showed a suppression of the superconducting gap of niobium electrodes and higher leakage currents. Both characteristics are presented in Fig. 2.18.

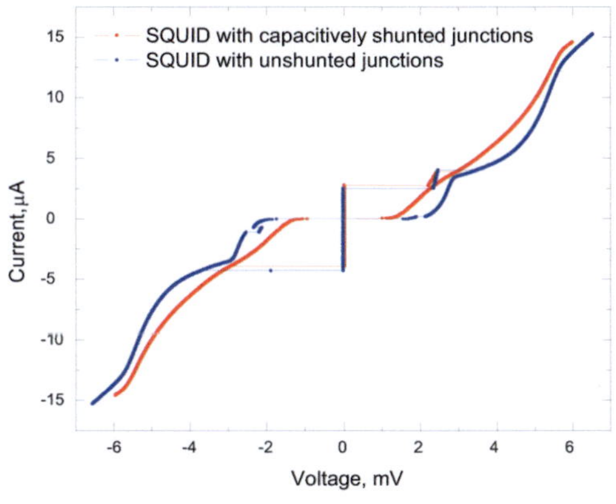

Figure 2.18: Current-voltage characteristics of the readout SQUIDs from the die D3 at 300 mK.

The chips from the wafer #3 showed poor adhesion of circuit contact pads. Pads were peeling off during bonding. Fig. 2.19 shows a bonded qubit circuit. There are several pads in the circuit which were almost destroyed by unsuccessful bonding attempts. When bonding it was hardly possible to distinguish whether only the top metallisation layer of the pad was destroyed or the whole pad.

2.7.2 Results

Dies from the wafer #3 were characterised in our Helium-3 cryostat at its base temperature.

Die D3

We measured a circuit with "medium" coupling between the redout SQUID and the qubit. The Figs. 2.20 and 2.21 show $I_c\, vs.\, I_s$ and $I_c\, vs.\, I_a$ dependences respectively. No feature which could be associated with the qubit switching was observed. However, during this experiment one of the branches of the antisymmetric control line was shunted to the ground. We have not found

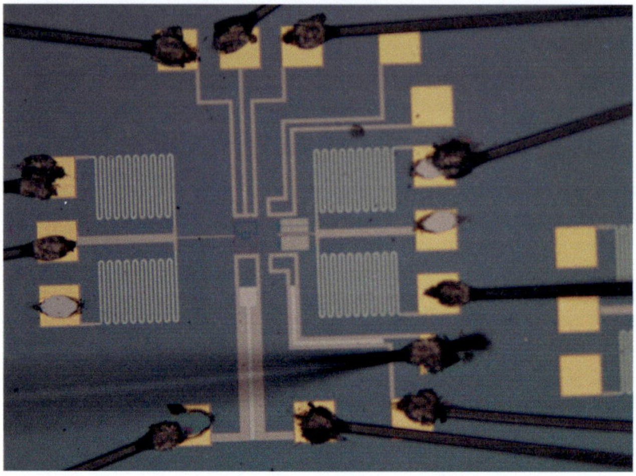

Figure 2.19: A photograph of a circuit from the chip D4 from the wafer #3.

any reason for this. Thus in this experiment the antisymmetric control line current I_a changed both symmetric and antisymmetric flux, *i.e.* both the hight of the potential barrier and the symmetry of the potential.

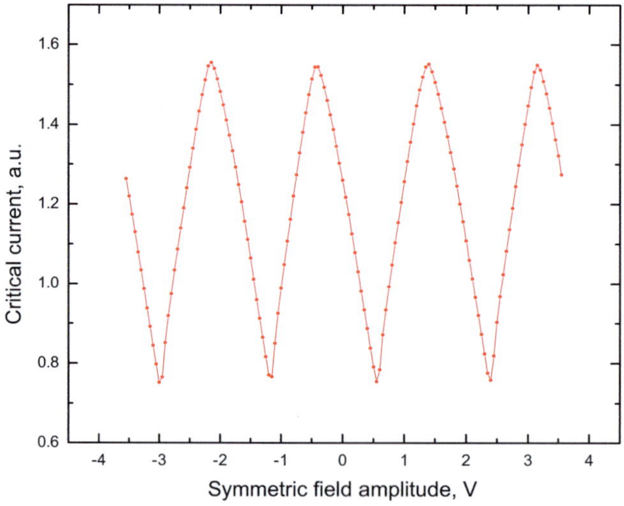

Figure 2.20: Dependence of the critical current of the readout SQUID on the chip D3 from the wafer #3 on symmetric control flux.

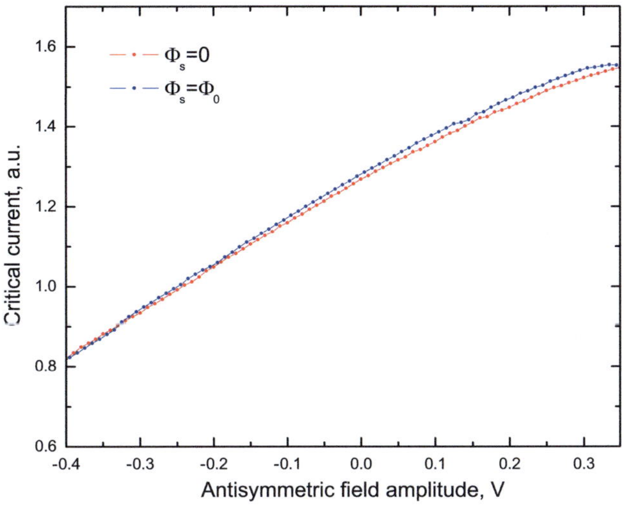

Figure 2.21: Dependence of the critical current of the readout SQUID on the chip D3 from the wafer #3 on antisymmetric control flux.

Die B4

We measured a circuit with "strong" coupling between the redout SQUID and the qubit. Due to the issue with contact pads the symmetric control line was disconnected throughout the experiment. We tried to produce both control fluxes passing different currents through the branches of the antisymmetric control line. The measured $I_c\, vs.\, I_s$ and $I_c\, vs.\, I_a$ dependences are presented in the Figs. 2.22 and 2.23 respectively.

In Fig. 2.23 one can readily see that two curves differ around zero antisymmetric control flux. The difference between two curves is presented in Fig. 2.24. We believe that the step at small positive values of the antisymmetric control flux is due to the qubit switching. The difference in flux between two qubit states is around $8\,\mathrm{m}\Phi_0$. This value is larger than calculated one of about $3\,\mathrm{m}\Phi_0$. It might be caused by larger than designed critical current of the outer qubit junctions.

When ramping antisymmetric control field in the opposite directions we saw a hysteretic behaviour of the $I_c\, vs.\, I_a$ dependences. An example of the hysteresis is presented in Fig. 2.25. We cannot explain this hysteresis, it might be caused by the current biasing scheme we used in this experiment.

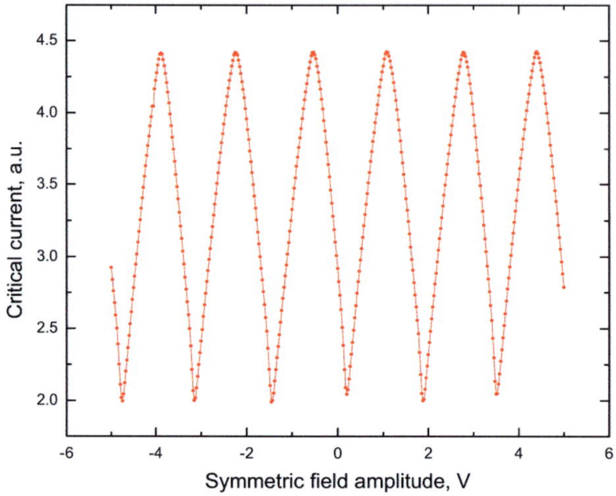

Figure 2.22: Dependence of the critical current of the readout SQUID on the chip B4 from the wafer #3 on symmetric control flux.

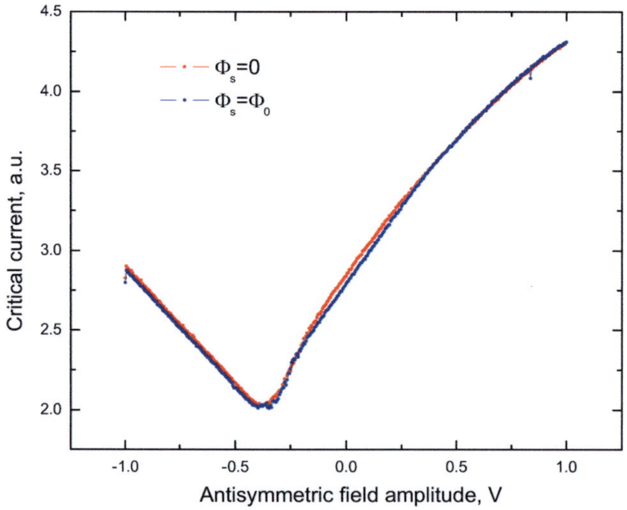

Figure 2.23: Dependence of the critical current of the readout SQUID on the chip B4 from the wafer #3 on antisymmetric control flux.

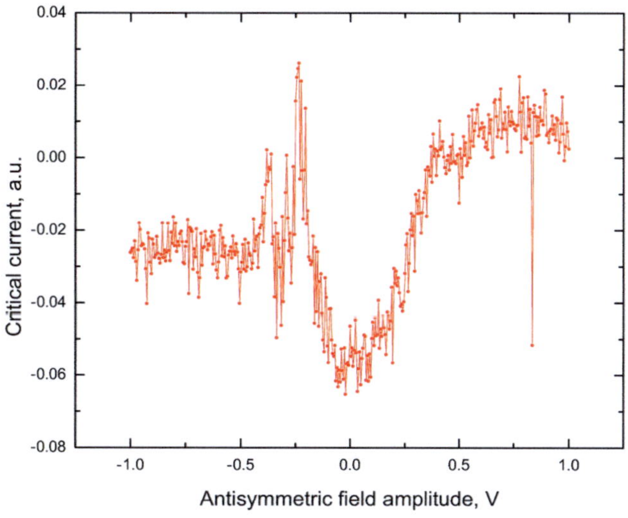

Figure 2.24: Difference between two curves presented in Fig. 2.23.

Die C4

We tried to measure a circuit with "medium" coupling between the redout SQUID and the qubit. Unfortunately the readout SQUID was broken.

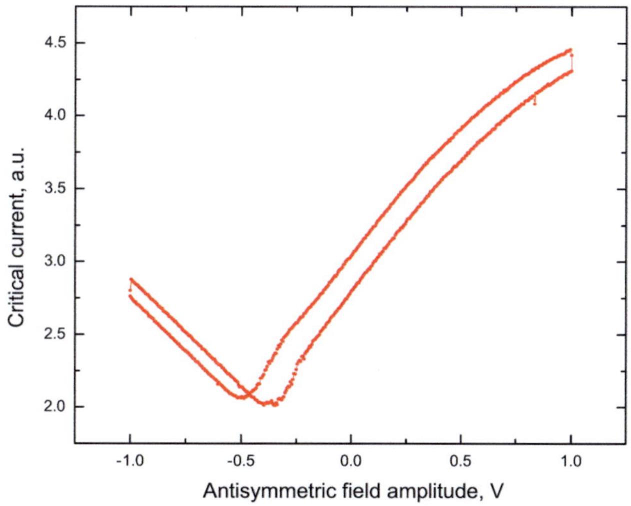

Figure 2.25: Hysteretic dependence of the critical current of the readout SQUID on the chip B4 from the wafer #3 on antisymmetric control flux.

Die D4

In the circuit with "strong" coupling between the readout SQUID and the qubit one of the SQUID contact pads peeled off completely during bonding procedure. We only found it out after cooling the chip down.

In the circuit with "medium" coupling between the qubit and the readout SQUID with capacitively shunted junctions $I_c\, vs.\, I_a$ dependences show a rapid change of the flux penetrating the SQUID loop. The dependance is shown in Fig. 2.26. The positing of this feature depends on the applied symmetric control flux. Such response of the circuit is abnormal for a qubit with designed parameters. $I_c\, vs.\, I_s$ dependence is presented in the Fig. 2.27.

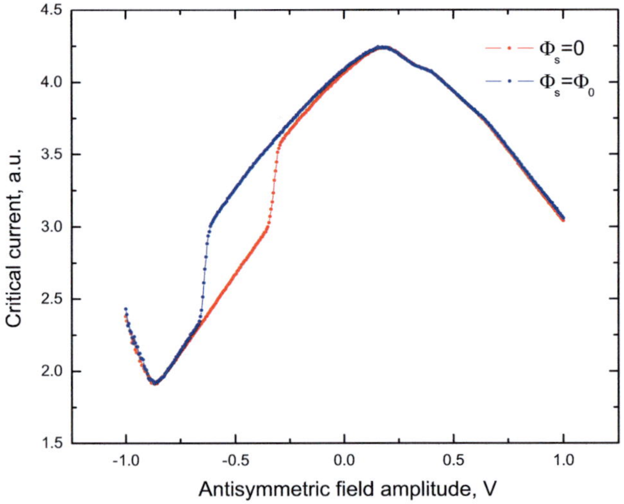

Figure 2.26: Dependence of the critical current of the readout SQUID on the chip D4 from the wafer #3 on antisymmetric control flux.

Die C5

We measured structures with "strong" and "medium" coupling between the qubit and the readout SQUID. The readout SQUID in the structure with "strong" coupling had capacitively shunted junctions.

$I_c\, vs.\, I_s$ dependence of the readout SQUID with "strong" coupling to the

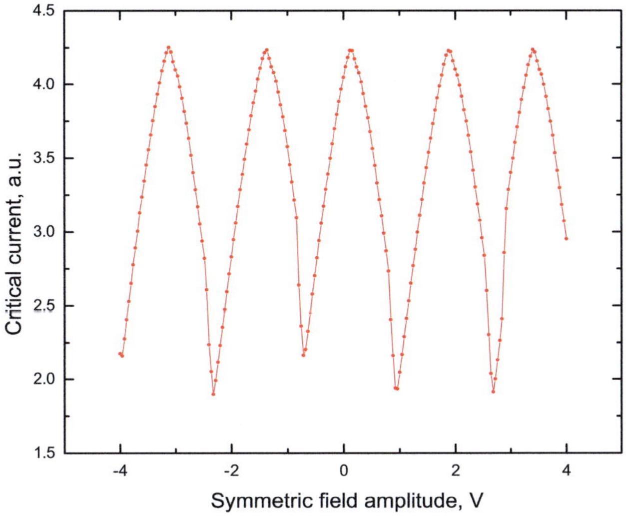

Figure 2.27: Dependence of the critical current of the readout SQUID on the chip D4 from the wafer #3 on symmetric control flux.

qubit is presented in Fig. 2.28.We could not see any change in the $I_c\, vs.\, I_a$ patterns (see Fig. 2.29) for this structure due to applied symmetric flux.

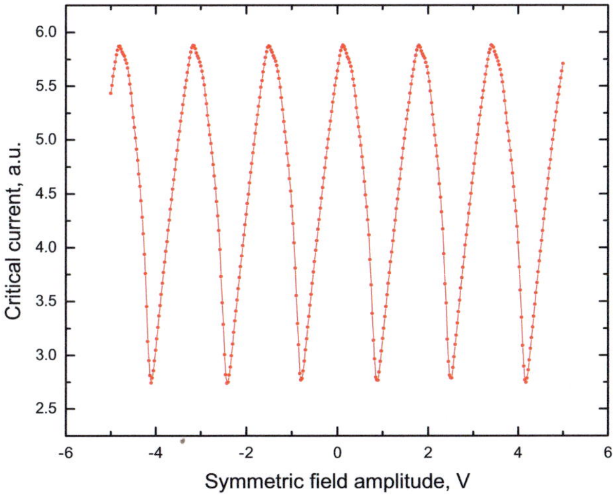

Figure 2.28: Dependence of the critical current of the readout SQUID with "strong" coupling to the qubit on the chip C5 from the wafer #3 on symmetric control flux.

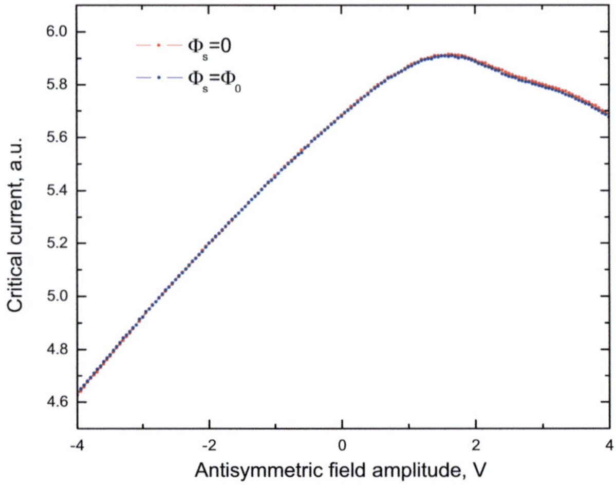

Figure 2.29: Dependence of the critical current of the readout SQUID with "strong" coupling to the qubit on the chip C5 from the wafer #3 on antisymmetric control flux.

The structure with "medium" coupling showed a qubit switching between the two classical states. $I_c\,vs.\,I_s$ and $I_c\,vs.\,I_a$ dependences are presented in Figs. 2.30 and 2.30 correspondingly. The S-curve is presented in Fig. 2.32. However the flux change in the SQUID loop induced by the qubit switching amounted to $23\,\mathrm{m}\Phi_0$, that is almost an order of magnitude larger than estimated value of $2.4\,\mathrm{m}\Phi_0$. It might be again explained by the fact, that the critical currents of the outer qubit junctions were larger than designed.

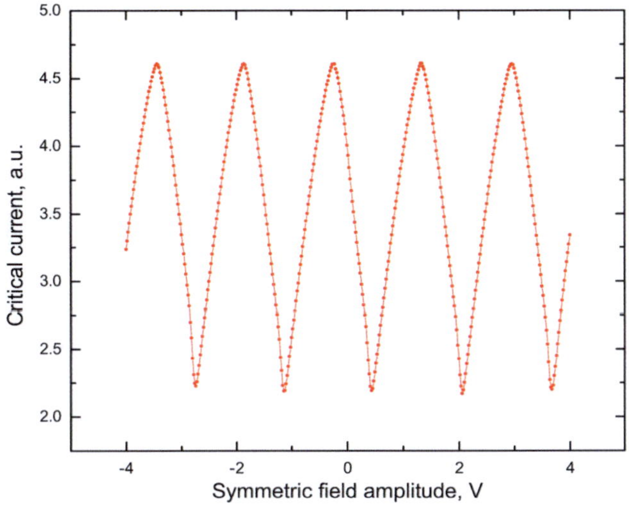

Figure 2.30: Dependence of the critical current of the readout SQUID with "medium" coupling to the qubit on the chip C5 from the wafer #3 on symmetric control flux.

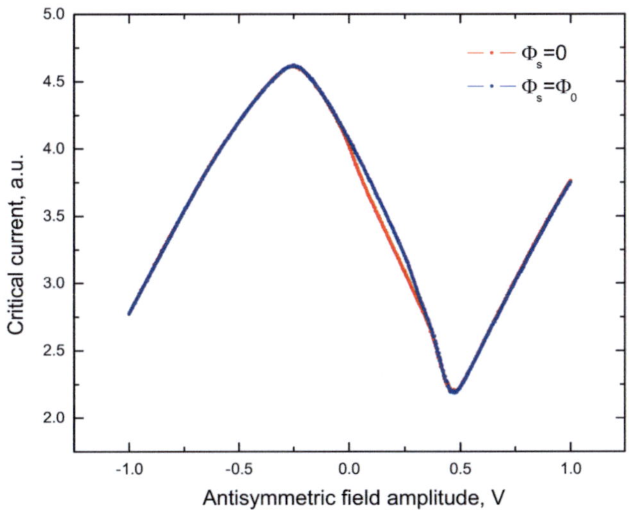

Figure 2.31: Dependence of the critical current of the readout SQUID with "medium" coupling to the qubit on the chip C5 from the wafer #3 on antisymmetric control flux.

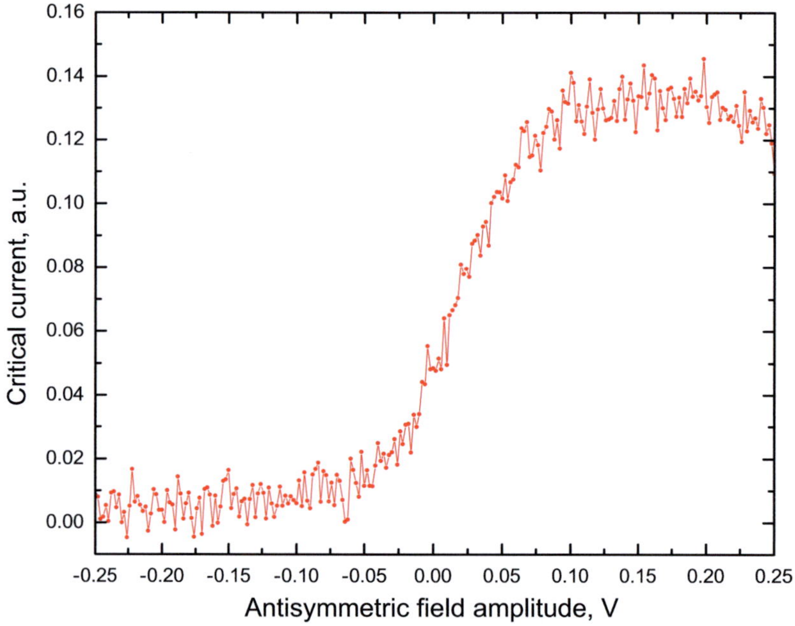

Figure 2.32: Difference between two curves presented in Fig. 2.31.

2.8 Summary

We designed our structures to realise the proposed "silent" two-cell qubits. At the operating point, the "silent" qubit should be insensitive to fluctuations caused by the flux control circuits.

We have observed switching of the two-cell flux qubit with a passive π-shifter between two classical states at 300 mK. The qubit switches around zero flux for both symmetric and antisymmetric flux components, which meets our expectation. This behaviour minimises the coupling of the qubit to the environment.

2.8.1 Wafers #1 and #2

The resolution of the experiment was limited due to the defects in the read-out SQUID junctions. Furthermore, the outer junctions in the qubits were damaged as well.

2.8.2 Wafer #3

Capacitive shunts did not help us to increase measurement resolution. The capacitors increased subgap leakage instead. The yield of smallest qubit junctions in our design was low. One of the reason for this might be the absence of the "fill" structures (metal structures used to improve the uniformity of the chemical-mechanical-polishing process) inside the readout SQUID in our design.

2.8.3 Outlook

Since yield of working structures produced by Lincoln Lab foundry depends on the topology of the circuits, optimisation process is quite difficult. Varying of the qubit topology instead of the coupling strength might give better result. The "fill" structures should be put all over the chip. Large area silicon dioxide capacitors should be avoided.

References

[1] Chiorescu, I., Nakamura, Y., Harmans, C. J. P. M. & Mooij, J. E. Coherent quantum dynamics of a superconducting flux qubit. *Science* **299,** 1869-1871 (2003).

[2] Mooij, J. E. *et al.* Josephson persistent-current qubit. *Science* **283,** 1036-1039 (1999).

[3] Yukon, S. P. A multi-Josephson junction qubit. *Physica C* **368,** 320-323 (2002).

[4] Yukon, S. P. Josephson junction triangular prism qubits coupled to a resonant LC bus. *Proceedings of MQC2 Conference, Naples* (2002).

[5] Majer, J. B., Butcher, J. R. & Mooij, J. E. Simple phase bias for superconducting circuits. *Appl. Phys. Lett.* **80,** 3638-3640 (2002).

[6] Bulaevskii, L. N., Kuzii, V. V. & Sobyanin, A. A. Superconducting system with weak coupling to the current in the ground state. *JETP Lett.* **25,** 290-294 (1977).

[7] Buzdin, A. I., Bulaevskij, L. N. & Panyukov, S. V. Critical-current oscillations as a function of the exchange field and thickness of the ferromagnetic metal (F) in an S-F-S Josephson junction. *JETP Lett.* **35,** 178-180 (1982).

[8] Dolan, G. J. Offset masks for lift-off photoprocessing. *Appl. Phys. Lett.* **31,** 337-339 (1977).

[9] Lisenfeld, J. *Experiments on superconducting Josephson phase quantum bits.* (PhD thesis, Friedrich-Alexander-Universität Erlangen-Nürnberg, 2008).

Chapter 3

Strong coupling of a quantum oscillator to a flux qubit at its symmetry point

A flux qubit biased at its symmetry point shows a minimum in the energy splitting (the gap), providing protection against flux noise. We have fabricated a qubit whose gap can be tuned fast and have coupled this qubit strongly to an LC oscillator. We show full spectroscopy of the qubit-resonator system and generate vacuum Rabi oscillations. When the gap is made equal to the oscillator frequency ν_{osc} we find the strongest qubit-resonator coupling ($g \sim 0.1\nu_{\rm osc}$). Here being at resonance coincides with the optimal coherence of the symmetry point. Significant further increase of the coupling is possible.*

*This experiment was performed in collaboration with A. Fedorov, P. Macha, P. Forn-Díaz, C. J. P. M. Harmans, and J. E. Mooij in the Quantum Transport group at the Kavli Institute of Nanoscience at Delft University of Technology.

3.1 Introduction

Superconducting qubits coupled to quantum oscillators have demonstrated a remarkable richness of physical phenomena in the last few years. After the first reports of coherent transfer and strong coupling [1, 2], we have witnessed a rapid development of the field called cavity or circuit quantum electrodynamics (cQED) using high quality superconducting oscillators in realising quantum gates [3], algorithms [4] as well as non-classical states of light and matter in artificially fabricated structures [5, 6]. Among the different superconducting implementations the transmon [2, 3, 4, 5] and the phase qubit [6] dominated this development. With flux qubits the avoided crossing between qubit and oscillator level was observed [7, 8, 9, 10] and the coherent single-photon exchange between qubit and oscillator at resonance was demonstrated [10]. In contrast to the transmon and the phase qubit, the coherence of the flux qubit is optimally preserved only in the symmetry point where the energy splitting is minimal ($h\Delta$) and do not easily coincide with frequency of an oscillator. We now have developed a flux qubit where the minimum energy splitting (the gap) can be tuned over a broad range on sub-ns time scales. This tunable-gap flux qubit can be kept at its high-coherence point while providing tunable and strong coupling; it even allows entering the ultra-strong coupling regime where the coupling strength approaches the qubit and/or oscillator energy [11, 12, 13, 14]. This regime is not easily accessible in quantum optics with atoms and ions. Here we demonstrate strong coupling combined with good coherence for the flux qubit and a lumped-element LC resonator, showing fast and long-living vacuum Rabi oscillations.

Superconducting qubits are fabricated objects and their relevant parameters, such as the energy (or frequency ν_{qb}) versus flux dependence and the coupling strength g, to a large extent can be chosen in the design phase. Typically, the qubit and oscillator frequencies ν_{qb} and ν_{osc} are in the GHz range. For strong coupling, where g exceeds the cavity and qubit loss rates, the rotating-wave approximation (RWA) can be applied and the system can be described by a Jaynes-Cummings type Hamiltonian. If g approaches the qubit or oscillator frequencies the RWA no longer holds, leading into the

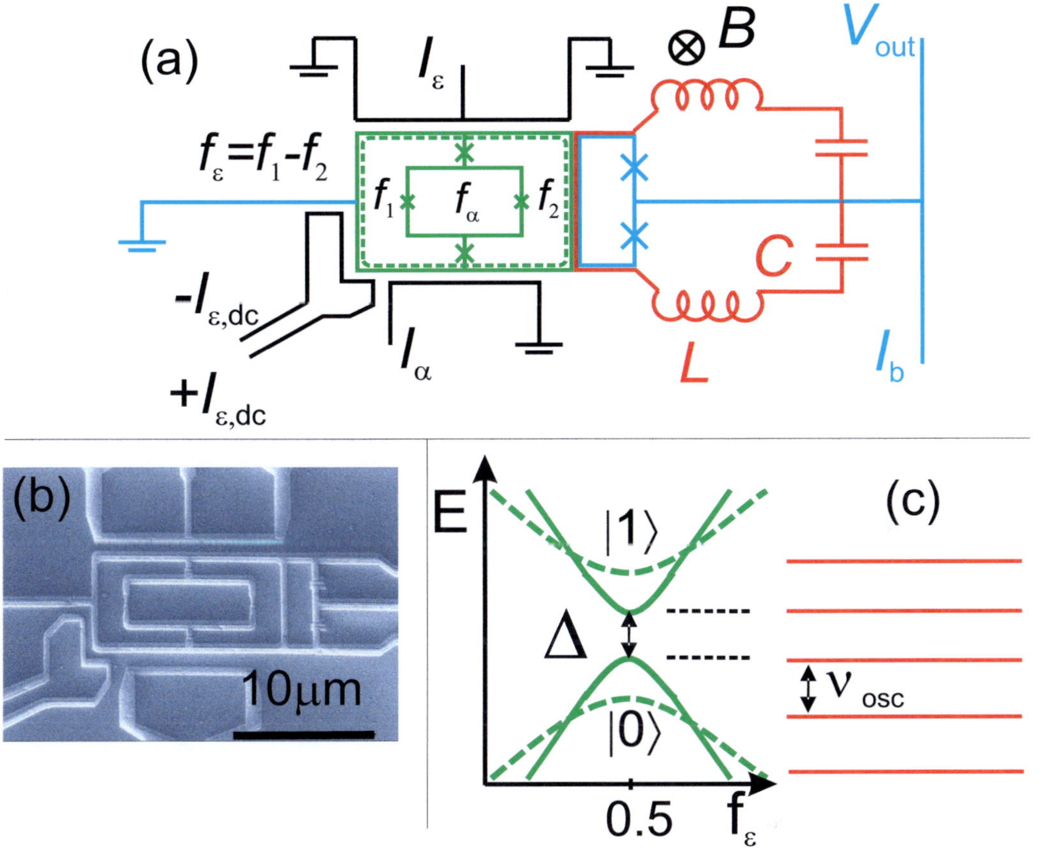

Figure 3.1: (a) Circuit schematics: the tunable gap flux qubit (green) coupled to a lumped element superconducting LC oscillator (red) and controlled by the bias lines I_ε,$I_{\varepsilon,dc}$ and I_α (black). SQUID (blue) measures the state of the qubit. The gradiometer loop (emphasised by a dashed line) is used to trap fluxoids. (b) Scanning Electron Micrograph (SEM) of the sample. (c) Energy diagram of the qubit-oscillator system. The minimum of energy of the qubit Δ is reached in the symmetry point when one fluxoid is traped in the gradiometer loop and the difference in magnetic fluxes $f_\varepsilon\Phi_0$ is 0 controlled by I_ε and $I_{\varepsilon,dc}$. By controlling the flux $f_\alpha\Phi_0$ with I_α and a uniform field B one can tune Δ in resonance with oscillator frequency $\nu_{\rm osc}$.

ultra-strong coupling regime [11, 12, 13, 14]. For a flux qubit the ratio $g/\nu_{\rm osc}$ can be made an order of magnitude stronger than for charge and phase qubits [15], while these latter devices have a coupling that can be several orders of magnitude larger than the atom-light interaction energy [2]. For good coherence, operating the qubit at its spectral symmetry point is required. So, experimentally combining galvanic coupling of oscillator and flux qubit with this symmetry point operation provides a major step forward in the development of cQED systems. For the flux qubit at the symmetry point the anharmonicity (distance between 2nd and 3rd level relative to qubit splitting) is very high, allowing very fast operation without quantum leakage.

3.2 Samples

Our tunable gap flux qubit is galvanically attached to a lumped-element LC resonator, as shown schematically in Fig. 3.1(a,b). A carefully imposed symmetric design complemented by suppressing thermal noise currents from low resistances in the various excitation lines resulted in good coherence. The flux qubit has a gradiometric topology [16] to suppress decoherence due to external flux noise, by having the Josephson junctions that form the qubit being symmetrically attached to the circumference loop; this loop is also employed to trap fluxoids (or 2π-phase-winding numbers) [17]. To obtain a variable-gap qubit the two center junctions form a SQUID structure, where the flux $f_\alpha \Phi_0$ *in situ* sets the effective critical current and in this way the qubit gap frequency Δ. It covers nearly two decades from 150 MHz to 12 GHz, providing full frequency control relative to the resonator at $\nu_{\mathrm{osc}} = 2.723$ GHz (see Fig. 3.1(c)). The Hamiltonian of the flux qubit can be written as

$$H_{\mathrm{qb}} = -h \left(\frac{\varepsilon(f_\varepsilon, f_\alpha)\sigma_z + \Delta(f_\alpha)\sigma_x}{2} \right),$$

where σ_z and σ_x are Pauli matrices written in the persistent current states basis; $h\varepsilon$ is the magnetic energy bias

$$h\varepsilon(f_\varepsilon, f_\alpha) = I_p(f_\alpha) f_\varepsilon \Phi_0,$$

with I_p being the circulating current in the qubit and $f_\varepsilon \Phi_0$ the magnetic flux in the qubit loop taken relative to $\Phi_0/2$. Note that the frustration

$2f_\varepsilon\Phi_0 = (f_1 - f_2)\Phi_0$ describes the difference in flux in the two loop halves of the gradiometer (Fig. 3.1(a)). Qubit excitation is obtained by the quadrupolar magnetic field generated by current in the symmetrically-split I_ε line, acting on the qubit flux $f_\varepsilon\Phi_0$. Similarly, the line I_α together with the homogeneous field B generated by an external superconducting coil only modulates $f_\alpha\Phi_0$ thus affecting the gap frequency Δ. Symmetry forbids crosstalk, thus implying a fully orthogonal control.

The qubit states are detected from measuring the qubit persistent current, using a DC-SQUID which is coupled to the qubit with a shared part of a wire of length $l = 6\,\mu$m, width $w = 350$ nm and thickness $t = 70$ nm leading to a mutual SQUID-qubit inductance $M \simeq 11$ pH. Half of M is provided by a kinetic inductance of the shared part

$$L_K \sim \frac{l}{tw}$$

which can be easily made even larger than the geometric contribution. At zero bias current the noise induced by the readout circuitry is decoupled in linear order both from f_α and f_ε due to the SQUID-qubit geometry. The junctions of the SQUID are shunted with two on-chip Al-AlO$_x$-Al parallel plate capacitors of $C = 4$ pF reducing the plasma frequency to $\nu_\mathrm{p} \approx 1.3$ GHz and providing additional filtering of high frequency noise from I_b.

The inductances of the superconducting lines L and capacitors C form a lumped element LC resonator with frequency

$$\nu_\mathrm{osc} = \frac{1}{2\pi\sqrt{2L\frac{C}{2}}} = 2.723\,\mathrm{GHz}.$$

The oscillator is coupled to the qubit via the same shared part of the wire with the qubit-SQUID mutual inductance $M \simeq 11$ pH (since M is much smaller than the Josephson inductances of the SQUID junctions most of the current flows through the shared part). Unlike the plasma resonance of the SQUID, the LC resonator contour does not include the Josephson junctions and the connections of resonator to the external circuit occurs in the voltage nodes of the resonance mode. Thus the resonator quality factor Q is not severely affected by the external impedance, reaching $Q \sim 6000$ for strong excitation and a few hundreds at low photon number. Being designed as part of the readout circuit, it was not optimised for high Q performance.

All structures excluding the bottom plate of the capacitors C were fabricated in the same layer of aluminum using standard lithography techniques. The bottom plates of the capacitors were fabricated in a separate layer followed by a plasma oxidation step resulting in a thin layer of Al-AlO$_x$-Al used as the dielectric of the capacitor.

3.3 Theoretical description

The interaction between the qubit and the oscillator can be described by

$$H_{\text{int}} = \frac{hg}{2}(a + a^\dagger)\sigma_z$$

written in the basis of the persistent current states, where $a^\dagger$, a are photon creation and annihilation operators of the oscillator defined in the oscillator Fock space $|n\rangle$. The qubit-oscillator interaction strength

$$hg = 2MI_pI_0,$$

where $M = 11$ pH is the qubit-oscillator mutual inductance, I_p is the qubit persistent current and

$$I_0 = \sqrt{\frac{h\nu_{\text{osc}}}{2L}} = 31 \text{ nA}$$

is the ground state current in the oscillator, with L the inductance of the resonator loop. In the energy eigenstates of the qubit, $|g\rangle$, $|e\rangle$, the system Hamiltonian reads

$$H = \frac{h\nu_{\text{qb}}}{2}\sigma_z + h\nu_{\text{osc}}\left(a^\dagger a + \frac{1}{2}\right) + \frac{hg}{2}\left(\cos\eta\,\sigma_z - \sin\eta\,\sigma_x\right)(a + a^\dagger), \quad (3.1)$$

where

$$\nu_{\text{qb}} \equiv \sqrt{\Delta^2 + \varepsilon^2}$$

is the splitting of the qubit energy levels and

$$\tan\eta \equiv \Delta/\varepsilon.$$

We perform spectroscopy for the two representative cases $\Delta = \nu_{\rm osc}$ and $\Delta < \nu_{\rm osc}$, both showing strong level repulsion. In particular in the former case we concurrently establish maximum coupling and maximum coherence, favoured by the coincidence of the resonance condition with the qubit symmetry point.

While the qubit is in the symmetry point, qubit and resonator coherently exchange the excitation with a frequency that is determined by the coupling and the detuning $\delta\nu = \Delta - \nu_{\rm osc}$ according to Ref. [18]

$$\nu_R = \sqrt{g^2 + \delta\nu^2}. \tag{3.2}$$

In case of nonzero magnetic energy bias ε, taking into account only $|0\rangle$, $|1\rangle$ oscillator states the Rabi frequency can be found analytically form (3.1) as

$$\nu_R = \left(g^2 + \nu_{\rm osc}^2 + \nu_{\rm qb}^2 - 2\sqrt{g^2\varepsilon^2 + \nu_{\rm qb}^2\nu_{\rm osc}^2}\right)^{1/2}. \tag{3.3}$$

Note that Eq. (3.3) reduces to Eq. (3.2) if $\varepsilon = 0$.

3.4 Measurement procedure

The experiment was conducted in a dilution refrigerator at a base temperature of 20 mK. The spectroscopy of the system was performed with the protocol sketched in Fig. 3.2. First we set the gap of the qubit with the external magnetic field B and apply a dc offset to $I_{dc,\varepsilon}$ to tune the qubit frequency to $\nu_{\rm qb} = 9$ GHz. In the second step we apply a square current pulse in I_ε, tuning the qubit splitting to the required frequency, combined with a microwave excitation. After each excitation pulse the qubit is returned to $\nu_{\rm qb} = 9$ GHz and a short bias current pulse I_b is applied to the SQUID for measurement of the qubit state.

By measuring the qubit away from its symmetry point we benefit from the high expectation value of the circulating current of the qubit eigenstates and a long relaxation time T_1, ranging from $T_1 \simeq 1\ \mu$s as a minimum in the symmetry point and gradually increasing to $T_1 > 4\ \mu$s at $\nu_{\rm qb} \simeq 9$ GHz with $\Delta \simeq 2$ GHz.

The sequence of operations to observe the vacuum Rabi oscillations starts by tuning the qubit gap Δ into the vicinity of $\nu_{\rm osc}$, setting f_α by the external

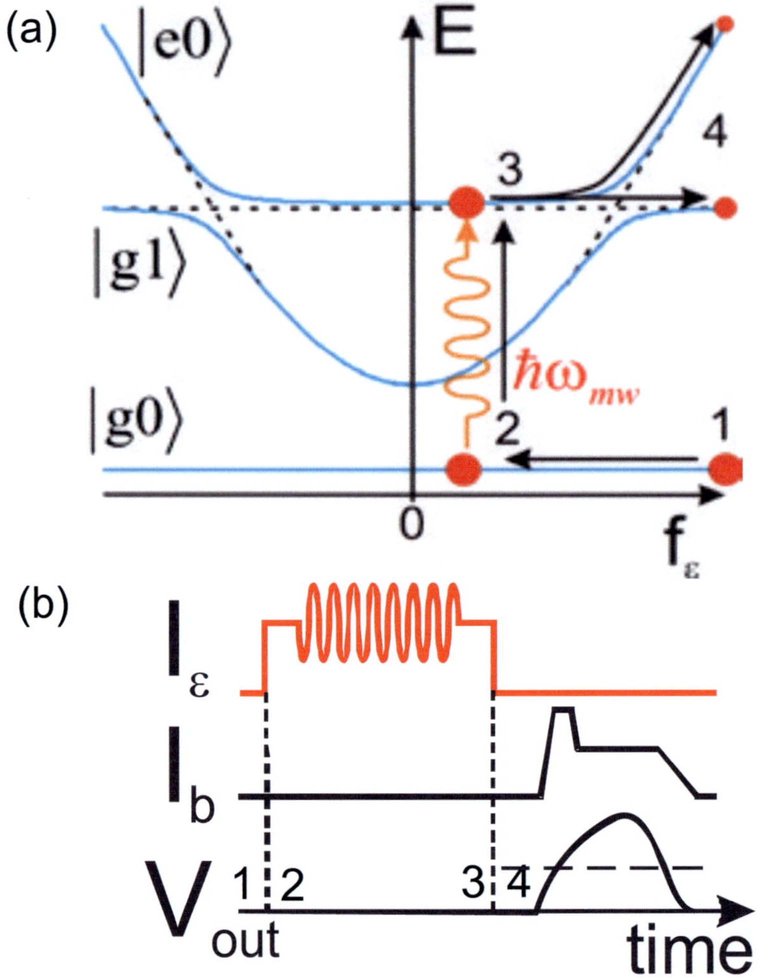

Figure 3.2: (a) During the spectroscopic measurement protocol the system undergoes Landau-Zener transitions transferring the excitation of the oscillator to the qubit. (b) Schematic representation of the control and measurement pulses.

magnetic field. The qubit is tuned to $\nu_{\rm qb} = 7$ GHz by f_ε , and a π-pulse is applied to excite the qubit. Subsequently the qubit is taken to the symmetry point by means of a fast 0.3 ns rise time pulse. As the qubit energy changes fast relative to the coupling strength g, this transfer is non-adiabatic. The qubit is kept here for a time Δt, returned fast to the 7 GHz level and read out. While the qubit is in the symmetry point, qubit and resonator coherently exchange the excitation with a frequency given by Eq. (3.2).

3.5 Experimental results

Fig. 3.3 shows a spectrum of the system for $\Delta < \nu_{\rm osc}$. In order to be resonant with the oscillator the qubit has to be tuned away from its symmetry point. The clear observation of the level anti-crossing confirms that the system is in the cQED regime. The size of the anti-crossing of 180 MHz is consistent with the design parameters combined with the independent measurement of g (see Fig. 3.5) and the interaction angle η.

To observe the spectral line of the resonator we use a Landau-Zener transition at the anti-crossing of the qubit and resonator energies. A passage through the anti-crossing region performed with a DC-shift pulse on I_ε with a rise time of 4 ns is found to lead to a $\sim 25\%$ probability of an oscillator photon to be converted to the excited state of the qubit. The latter can be detected by the SQUID and the resonator line becomes visible on a spectrum.

To spectroscopically observe the level anti-crossing one also has to be able to excite the qubit strongly by microwave radiation without appreciably exciting the resonator [19]. In our design (Fig. 3.1(a)) the I_ε current distribution has a quadruple symmetry with respect to the trap loop, in this way effectively decoupling it from the LC circuit.

The vacuum Rabi oscillations are shown in Fig. 3.4(a). For each value of f_α (and therefore Δ), the probability to find the qubit in one of its eigenstates oscillates as a function of Δt with a frequency that is minimal for $f_\alpha \cong -0.202$, the point where $\Delta = \nu_{\rm osc} = 2.723$ GHz . Fig. 3.4(b) shows the spectrum as a function of f_α, with the avoided crossing clearly visible. From the slope the value $d\Delta/df_\alpha \approx 69.5$ GHz/mΦ_0 can be determined, which is used to estimate $\delta\nu$ as $\delta\nu = (d\Delta/df_\alpha)\delta f_\alpha$. By fitting to Eq. (3.2) the bare

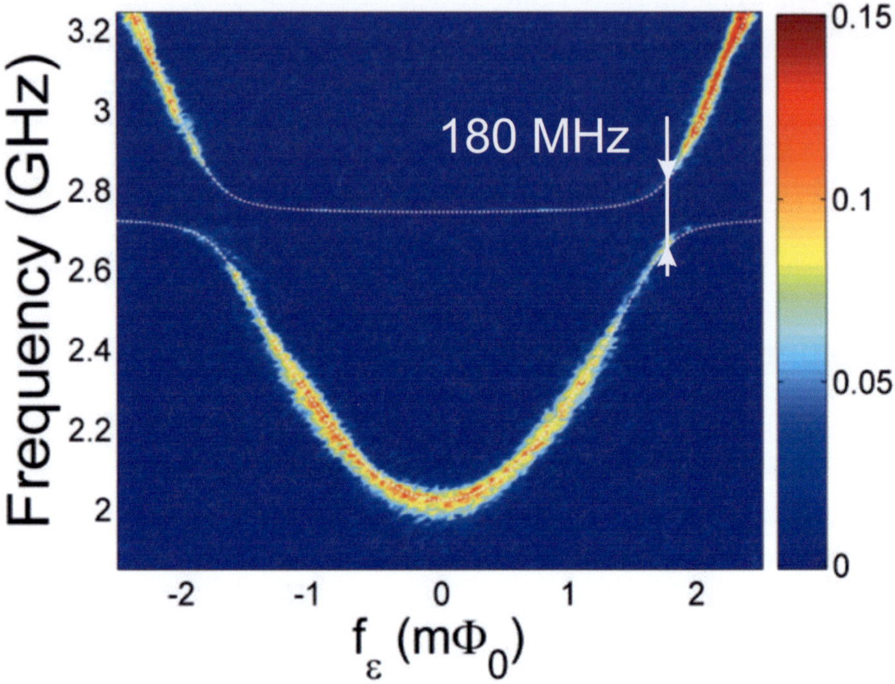

Figure 3.3: Microwave frequency vs magnetic frustration f_ε (controlled by the amplitude of the current pulse I_ε). The colour indicates the switching probability of the SQUID minus 0.5. The white dotted line is obtained from Eq. (3.1) with g = 239 MHz, Δ = 2.04 GHz, I_p = 250 nA. The anti-crossing size of 180 MHz corresponds to the qubit-oscillator coupling strength reduced by $\sin\eta$.

coupling g is found to be 239 MHz (see Fig. 3.5).

We now focus on the most interesting on-resonance regime with $\Delta = \nu_{\text{osc}}$. From (3.1) one can see that here the qubit-oscillator coupling is fully transversal making the effective coupling $g \sin\eta$ to attain its maximum value g. The measurement of the corresponding spectrum, shown in Fig. 3.6(a), indeed exhibits the maximum anti-crossing of 239 MHz corresponding to the highest photon exchange rate between the oscillator and the qubit.

Interestingly Landau-Zener transitions change now qualitatively: going through the anti-crossing the qubit and the oscillator almost fully exchange their populations creating strong asymmetry in the visibility of the spectral

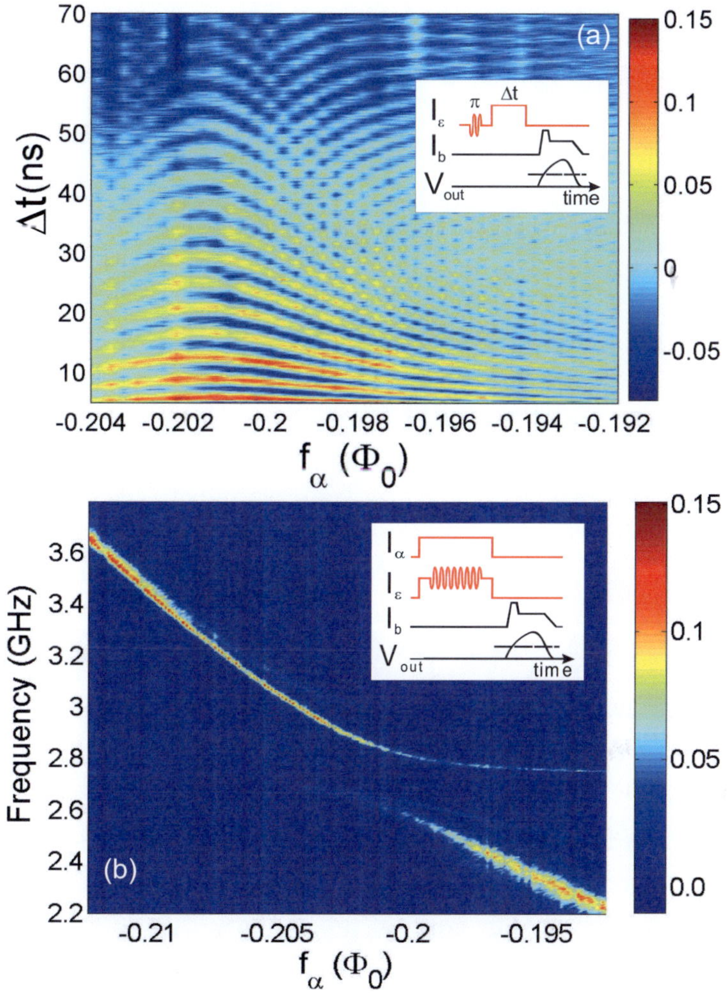

Figure 3.4: Vacuum Rabi oscillations (a) and microwave frequency (b) vs magnetic frustration f_α. In the experiment the qubit was kept in its symmetry point ($\varepsilon = 0$) by appropriately adjusting the amplitude of the current pulse I_ε while Δ was changed by f_α with use of the external magnetic field B (a) or by applying the current pulse I_α for fixed B (b). The colour scale in (a,b) shows the switching probability of the SQUID minus 0.5.

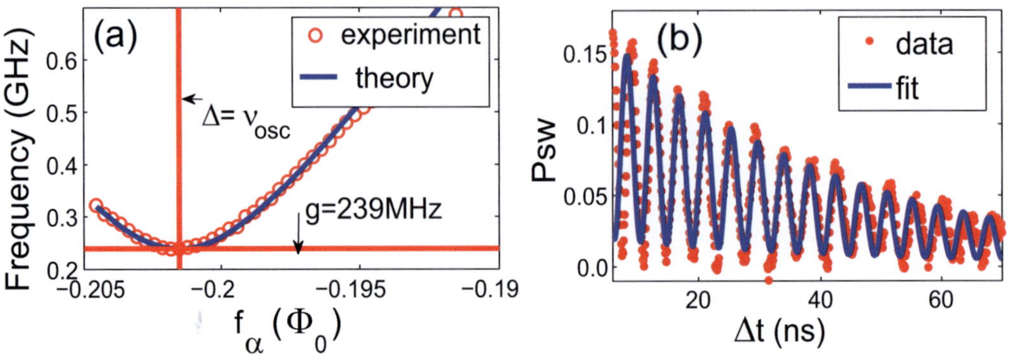

Figure 3.5: (a) Frequency of the vacuum Rabi oscillations extracted from data shown in Fig. 3.4(a) and theoretical estimation (blue line) from Eq. (3.2) as a function of f_α. The minimum in the oscillations frequency determines the bare qubit-oscillator coupling $g = 239$ MHz and corresponds to the resonance conditions $\Delta = \nu_{\rm osc} = 2.723$ GHz. (b) Single trace of the vacuum Rabi oscillations for $\Delta \simeq \nu_{\rm osc}$.

lines in Fig. 3.6(a) (for $\Delta < \nu_{\rm osc}$ (see Fig. 3.3) the qubit and the oscillator tend to retain their populations). This effect was used by Koch *et al.* [20] to create coherent operations.

In Figure 3.6(b) we demonstrate vacuum Rabi oscillations for different magnetic frustrations f_ε. The Rabi frequency in this case is given by Eq. (3.3), which explains the measured data as shown in Fig. 3.6(c).

Fig. 3.6(b) shows another interesting feature: ν_R is constant in a large range around $f_\varepsilon \sim 0$ due to strong qubit-oscillator interaction. Consequently the qubit-oscillator photon exchange becomes highly robust against the flux noise and f_ε control imperfections.

3.6 Discussion

Implementation of the gap control loop might in principle lead to extra decoherence. However, in practice the effect of flux noise in α-loop in our design is estimated to be about two orders of magnitude smaller than that of the

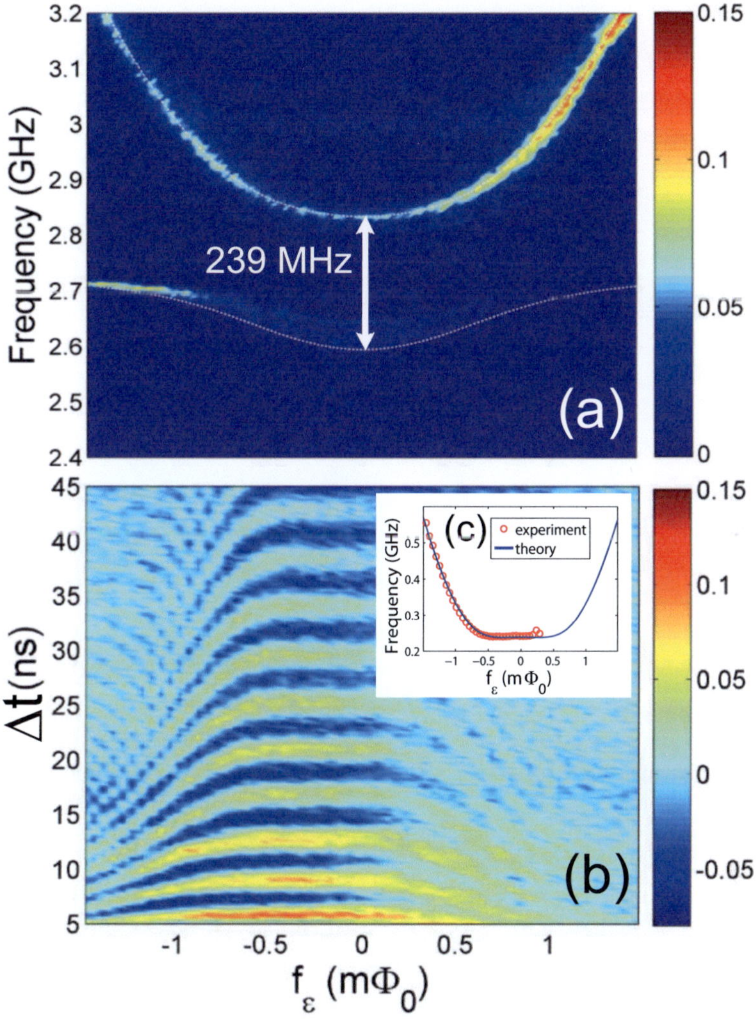

Figure 3.6: (a) Microwave frequency vs magnetic frustration f_ε . The dotted white line is obtained from Eq. (3.1) with $g = 239$ MHz, $I_p = 232$ nA and $\Delta = \nu_{\mathrm{osc}} = 2.723$ GHz. The observed anti-crossing is maximal due to fully transverse coupling of the qubit to the oscillator $\eta = \pi/2$. (b) Vacuum Rabi oscillations for different values of f_ε. In the experiment f_ε was controlled by the amplitude of the current pulse I_ε while Δ was tuned to ν_{osc} by changing the external magnetic field B. The inset (c) shows frequency of vacuum Rabi oscillations extracted from data (red circles) and estimated from Eq. (3.3) (blue line). The colour indicates the switching probability of the SQUID minus 0.5.

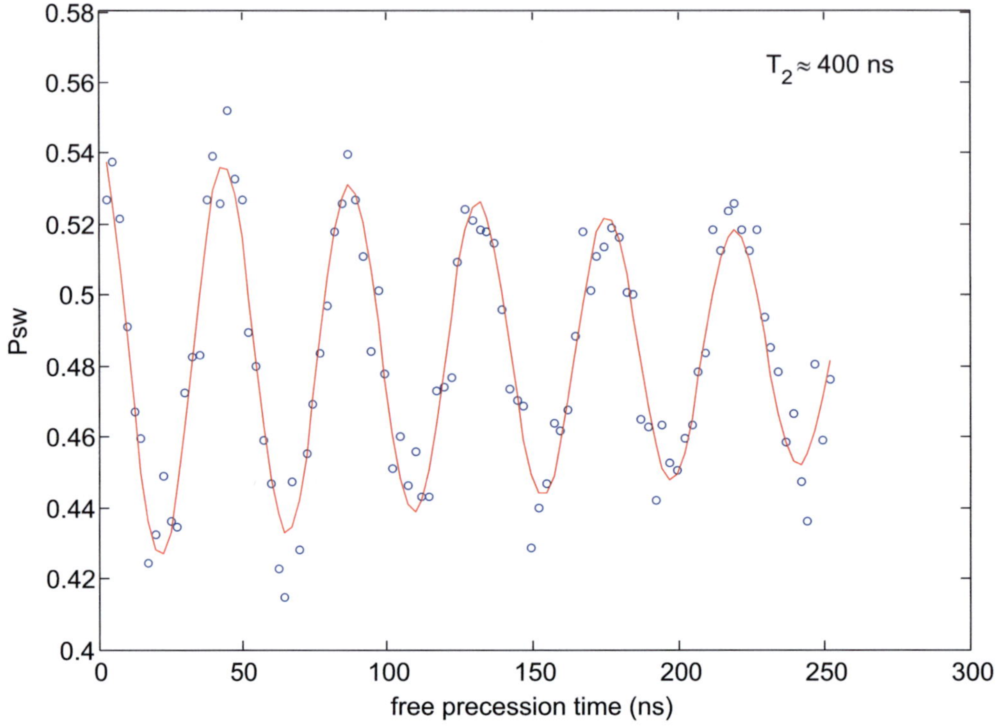

Figure 3.7: Ramsey oscillations. Red line is a fit to a cosine function decaying exponentially with time constant $T_2 \simeq 400$ ns.

ε-loop. Measuring the qubit in the symmetry point we find decay time of Ramsey oscillations $T_2 \simeq 400$ ns (see fig. 3.7) and $T_1 \simeq T_{\mathrm{Rabi}} \simeq 1\,\mu$s for $\Delta \sim 1.5 - 6$ GHz (see figs. 3.8,3.9). While T_1 and T_{Rabi} are in accordance with design values we observed no dependence of $T_2 < 2\,T_1$ on Δ which rules out the flux noise in both f_ε, f_α as a limiting decoherence source in the symmetry point. Detailed analysis of the data showed that the limitation of T_2 is most likely not due to the gap or the epsilon controls but due to a low frequency interference with the dilution refrigerator cooling cycle. Further measurements are required to verify the intrinsic value of T_2.

Since the qubit is optimally protected from low-frequency flux noise in the symmetry point $f_\varepsilon = 0$ the vacuum Rabi oscillations show the longest decay time of ~ 40 ns. This is limited only by the losses in the resonator,

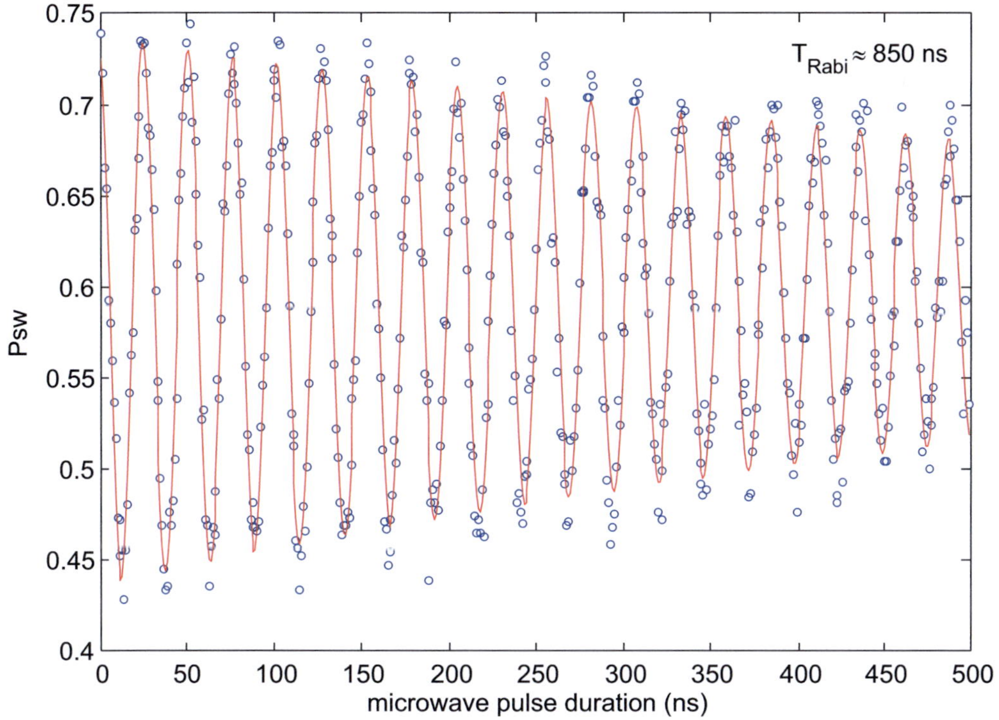

Figure 3.8: Rabi oscillations at the qubit symmetry point. Red line is a fit to a cosine function decaying exponentially with time constant $T_{\text{Rabi}} \simeq 850$ ns.

as measured coherence times in the symmetry point are much longer. Out of the symmetry point we measure the usual rapid degradation of the qubit coherence to $T_2 \sim 15 - 20$ ns for $\varepsilon \gg \Delta$ [21, 22] due to flux noise which precludes generation of long-living vacuum Rabi oscillations. Obviously, by using a fully compatible fabrication technology optimised for high Q resonators [2, 3, 4, 5] it is possible to achieve a ratio of

$$\frac{g}{\max\left(T_2^{-1}, T_1^{-1}, 2\pi\nu_{\text{osc}}Q^{-1}\right)} > 100$$

necessary for creating qubit-oscillator entanglement with high fidelity.

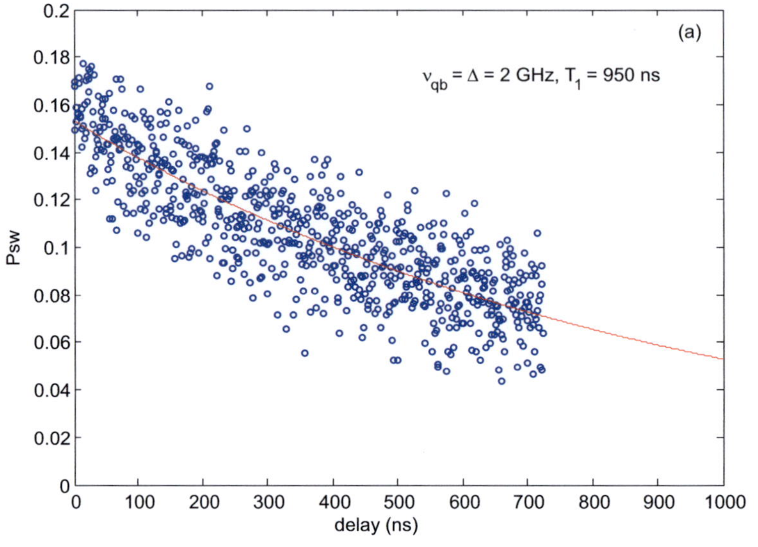

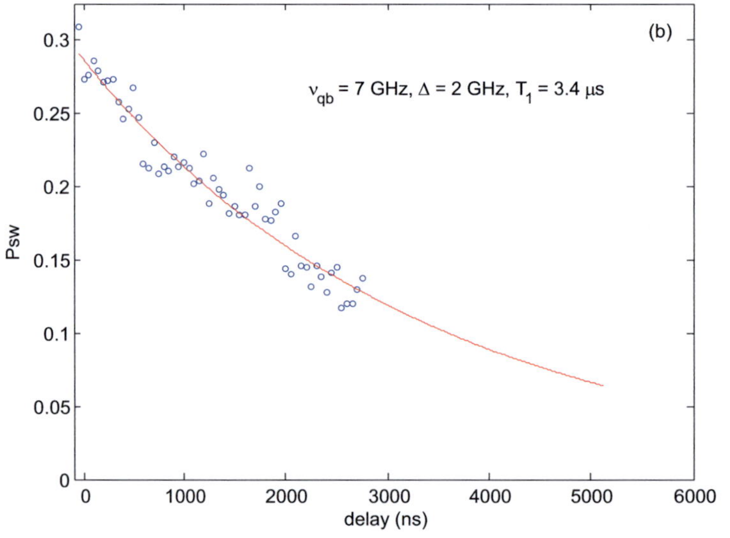

Figure 3.9: Decay of the qubit excited state. Red line is a fit to an exponential decay with time constant (a) $T_1 \simeq 950$ ns for the qubit at the symmetry point and (b) $T_1 \simeq 3.4\,\mu$s for the qubit magnetically biased to $\nu_{\rm qb} \approx 7$ GHz. The qubit gap is 2 GHz. T_1 is gradually increasing with magnetic bias energy $h\varepsilon$ as expected.

3.7 Summary

We experimentally studied the tunable gap flux qubit coupled galvanically to a superconducting lumped-element LC resonator. We measured the avoided level crossings and generated vacuum Rabi oscillations for two representative cases: the gap was tuned substantially below and in resonance with the resonator frequency. For the particularly interesting case of $\Delta = \nu_{\rm osc}$ the qubit reaches the resonance condition in its symmetry point thus combining two most desired ingredients of the cQED regime: strong coupling and optimal coherence. Here the avoided level crossing attains its maximal value of $g \simeq 0.09 \times \nu_{\rm osc}$ and at the same time the qubit is effectively protected from 1/f flux noise resulting in the longest and fastest sequence of on-resonant vacuum Rabi oscillations. We emphasise that with the galvanic coupling the interaction strength can be readily increased by use of simple design changes, reaching the ultra-strong coupling regime $g \sim \{\nu_{\rm osc},\, \nu_{\rm qb}\}$.

References

[1] Chiorescu, I. *et al.* Coherent dynamics of a flux qubit coupled to a harmonic oscillator. *Nature* **431,** 159-162 (2004).

[2] Wallraff, A. *et al.* Strong coupling of a single photon to a superconducting qubit using circuit quantum electrodynamics. *Nature* **431,** 162-167 (2004).

[3] Majer, J. *et al.* Coupling superconducting qubits via a cavity bus. *Nature* **449,** 443-447 (2007).

[4] DiCarlo, L. *et al.* Demonstration of two-qubit algorithms with a superconducting quantum processor. *Nature* **460,** 240-244 (2009).

[5] Fink, J. M. *et al.* Climbing the Jaynes-Cummings ladder and observing its square root of n nonlinearity in a cavity QED system. *Nature* **454,** 315-318 (2008).

[6] Hofheinz, M. *et al.* Synthesizing arbitrary quantum states in a superconducting resonator. *Nature* **459,** 546-549 (2009).

[7] Il'ichev, E. *et al.* Continuous monitoring of Rabi oscillations in a Josephson flux qubit. *Phys. Rev. Lett.* **91,** 097906 (2003).

[8] Abdumalikov, A. A., Astafiev, O., Nakamura, Y., Pashkin, Y. A. & Tsai, J. S. Vacuum Rabi splitting due to strong coupling of a flux qubit and a coplanar-waveguide resonator. *Phys. Rev. B* **78,** 180502 (2008).

[9] Deppe, F. *et al.* Two-photon probe of the Jaynes-Cummings model and controlled symmetry breaking in circuit QED. *Nature Phys.* **4,** 686-691 (2008).

[10] Johansson, J. *et al.* Vacuum Rabi oscillations in a macroscopic superconducting qubit LC oscillator system. *Phys. Rev. Lett.* **96,** 127006 (2006).

[11] Bourassa, J. *et al.* Ultrastrong coupling regime of cavity QED with phase-biased flux qubits. *Phys. Rev. A* **80,** 032109 (2009).

[12] Peropadre, B., Forn-Díaz, P., Solano, E. & Garcia-Ripoll, J. J. Switchable ultrastrong coupling in circuit QED. arXiv:0912.3456 (2009).

[13] Ashhab, S. & Nori, F. Qubit-oscillator systems in the ultrastrong-coupling regime and their potential for preparing nonclassical states. arXiv:0912.4888 (2009).

[14] Niemczyk, T. *et al.* Beyond the Jaynes-Cummings model: circuit QED in the ultrastrong coupling regime. arXiv:1003.2376 (2010).

[15] Assuming galvanic and capacitive couplings for the flux and the phase qubits, repspectively, ratio $g/\nu_{\rm osc}$ for the flux qubit can be $\langle\hat{n}\rangle\Phi_0 I_{\rm rms}/(2eV_{\rm rms}) = 0.25R_K/Z_0 \approx 10$ where e is a charge of an electron, R_K a quantum of resistance, Z_0 is the characteristic impedance of the resonator and $\langle\hat{n}\rangle$ is a number of charge states.

[16] Paauw, F. G., Fedorov, A., Harmans, C. J. P. M. & Mooij, J. E. Tuning the gap of a superconducting flux qubit. *Phys. Rev. Lett.* **102,** 090501 (2009).

[17] Majer, J. B., Butcher, J. R. & Mooij, J. E. Simple phase bias for superconducting circuits. *Appl. Phys. Lett.* **80,** 3638-3640 (2002).

[18] Yamamoto, Y. & Imamoglu, A. *Mesoscopic Quantum Optics* (John Wiley and Sons, New York, 1999).

[19] Fink, J. M. *et al.* Thermal excitation of multi-photon dressed states in circuit quantum electrodynamics. *Physica Scripta* **T137,** 014013 (2009).

[20] Koch, R. H. *et al.* Experimental demonstration of an oscillator stabilized Josephson flux qubit. *Phys. Rev. Lett.* **96,** 127001 (2006).

[21] Yoshihara, F., Harrabi, K., Niskanen, A. O., Nakamura, Y. & Tsai, J. S. Decoherence of flux qubits due to 1/f flux noise. *Phys. Rev. Lett.* **97,** 167001 (2006).

[22] Kakuyanagi, K. *et al.* Dephasing of a superconducting flux qubit. *Phys. Rev. Lett.* **98,** 047004 (2007).

Chapter 4

Implementation of superconductor-ferromagnet-superconductor π-shifters in superconducting digital and quantum circuits

High operation speed and low energy consumption may allow the superconducting digital single flux quantum circuits to outperform traditional complementary metal-oxide-semiconductor logic. The remaining major obstacle to a high density of elements on a chip is a relatively large cell size necessary to hold a magnetic flux quantum Φ_0. Inserting a π-type Josephson junction [1, 2] in the cell is equivalent to applying flux $\Phi_0/2$ and thus makes it possible to solve this problem [3]. Moreover, using π-junctions in superconducting qubits may help to protect them from noise [4, 5]. Here we demonstrate the operation of two superconducting circuits – one of them is classical and one quantum – which both utilise such π-phase shifters realised using superconductor-ferromagnet-superconductor sandwich technology [6]. The classical circuit is based on single-flux-quantum cells, which are shown to be scalable and compatible with conventional niobium-based superconducting electronics. The quantum circuit is a π-biased phase qubit, for which we observe coherent Rabi oscillations. We find no degradation of the measured coherence time compared to that of a reference qubit without π-junction.

4.1 Introduction

In superconducting circuits, currents can flow without applying any electric field. The role of the electrostatic potential difference required to drive a current in conventional circuits is played here by a difference φ between the phases of the superconducting order parameters. In the absence of current, φ is zero, but this can be altered by inserting a particular type of superconducting weak link, a so-called π-junction [1, 2] yielding a phase shift of π. The fundamental property of superconducting weak links is a 2π-periodic current-phase relation. The supercurrent through a conventional Josephson junction is usually described by the harmonic relation $I_s = I_c \sin \varphi$, where I_c is the critical current, while the π-junction has the inverse current-phase relation $I_s = I_c \sin(\varphi + \pi) = -I_c \sin \varphi$. The π-junctions were theoretically proposed about three decades ago, whereas their remarkable properties have been demonstrated in experiments notably later [7, 8, 6]. Practical implementations of π-junctions have been widely discussed for a variety of different technologies. These include approaches using superconductors with *d*-wave order parameter symmetry [7, 9, 10], circuits with nonequilibrium current injection [8], junctions with ferromagnetic layers [6], and junctions with gated carbon nanotubes [11].

The ideas of using π-junctions in superconducting classical and quantum circuits have been explored in several theoretical proposals. In classical digital logic, a complementary Josephson junction inverter [12] was suggested as a superconducting analog of the complementary metal-oxide-semiconductor (CMOS) logic. It relies on using Superconducting Quantum Interference Devices (SQUIDs) of conventional (0−junctions) and π-types and requires similar parameters as I_c and normal state resistance for 0- and π-junctions. These technologically stringent requirements can be softened by using an alternative "asymmetric" approach [3] which employs π-junctions as passive phase shifters (phase inverters) in basic cells of the modified Single-Flux Quantum (SFQ) logic. Here the π-junction critical current I_c is chosen to be much larger than that of conventional 0−junctions employed in the very same SFQ cell, so the phase difference across the π-junction is always close to π even at zero magnetic field. Since the total change of the order parameter's phase over the closed path must become a multiple of 2π, the 'missing' phase difference of π or $-\pi$ is induced on the rest part of the cell by a spontaneously generated superconducting current.

The first proposal for using a loop with an integrated π-junction as a superconducting quantum circuit [4, 5] featured a superposition of two persistent current states in a loop at zero magnetic field, in analogy to a spin-1/2 system. The π-junctions required here must have very low dissipation (high normal resistance), which so far has seemed unattainable for any of the existing technologies for making π-junctions. The alternative usage of π-junctions as passive phase shifters offers an advantage for the operation of superconducting flux qubits at the degeneracy point requiring zero or a very small external magnetic field. Potentially, this allows noise and electromagnetic interference induced by magnetic field sources to be minimised. There remains an open question: Do π-junctions themselves introduce any intrinsic decoherence when they are inserted into a superconduction quantum circuit?

4.2 SFS π-junctions

The origin of the π-state in an SFS junction is an oscillating and sign-reversing superconducting order parameter in the ferromagnet close to the SF interface [2, 13]. Due to these oscillations, different signs of the order parameter can occur at the two banks of the SFS sandwich when the F-layer thickness is of the order of half an oscillation period, which corresponds to a sign change of the supercurrent and a negative Josephson coupling energy. This behaviour was first observed experimentally on Nb-CuNi-Nb sandwiches in Ref. [6]. Further experiments reported the spontaneous flux [14] and half-periodical shifts of the superconducting interferometer $I_c(H)$ dependence [15] as well as a sign change of the junction current-phase relation [16]. Recently, the critical current density of the Nb-$Cu_{0.47}Ni_{0.53}$-Nb π-junctions was pushed above 1000 A/cm^2 [17]. These junctions are compatible with conventional niobium thin film technology and thus can be easily integrated in the conventional fabrication process of superconducting digital circuits.

4.3 Samples

Details on the fabrication technique for tunnel Nb/Al/AlO_x/Nb junctions employed in complementary SQUID circuits are presented in Ref. [18]. In brief, a three-layer Nb/Al/AlO_x/Nb structure is deposited by magnetron sputtering. The layers have thicknesses of 180, 7, and 80 nm, respectively. Aluminum is oxidised in pure oxygen to form a tunnel barrier having a critical current density of about 200 A/cm^2. The junction area, here 10 μm^2, is defined by reactive ion etching and subsequent SiO_2 deposition. Resistive shunts in parallel to the tunnel junctions are formed by a molybdenum layer with a specific resistance of 2 Ω per square.

For fabrication of SFS π-junctions, the bottom Nb-electrode with thickness of 110 nm was fabricated by dc-magnetron sputtering followed by a lift-off process. A 15 nm thick $Cu_{0.47}Ni_{0.53}$ layer (F layer) was deposited by rf sputtering after ion cleaning of the niobium surface. Afterwards, the insulating layer having a window which determines the junction area was prepared by the lift-off process. We used a 150 nm thick SiO film as insulator, which was thermally evaporated. The fabrication procedure was completed by Ar plasma cleaning and dc-magnetron sputtering of the upper niobium electrode of 240 nm thickness. A 10×10 μm^2 junction normal resistance R_n is about 150 $\mu\Omega$. The critical currents of such π-junctions are about 200 μA and hence the junctions do not switch to the resistive state when embedded in loops with conventional tunnel junctions having critical currents of about 10 μA. This large difference between two critical currents means that during the dynamic switchings in the rest of the circuit, π-junctions do not introduce any noticeable phase shifts deviating from π.

The qubit circuit was fabricated in a standard Nb/Al-AlO_x/Nb trilayer process, whereas the π-junction was integrated subsequently by performing the further lithographic steps described above.

4.4 Realisation of complementary Josephson junction logic cells using a passive π-shifter

To verify the operation of π-junction phase shifters in an analog regime, we fabricated two geometrically identical superconducting loops on a single Si substrate (see Fig. 4.1(c)) schematically shown in Figs. 4.1(a) and (b). Circuit (b) is a two-junction interferometer conventionally called a DC-SQUID. Configuration of circuit (a) is nominally identical to (b), except that an SFS π-junction has been additionally inserted in the left branch of the loop, seen in the lower left corner of the circuit image in Fig. 4.1(c). The on-chip distance between the centers of the two loops is 140 μm, so both interferometers are exposed to the same magnetic field during the experiment. The π-junction critical current is much larger than one of the tunnel junctions. Therefore, during the dynamic switchings in the rest of the circuit, π-junctions do not introduce any noticeable phase shifts deviating from π.

The dependencies of the critical currents $I_c(H)$ of the two devices shown in Figs. 4.1(a) and (b) are presented in Fig. 4.2. Whereas both curves have the same shape, they are shifted by a half-period. A small offset of the symmetry axes for both curves from the zero-field value is due to a residual magnetic field in the cryostat. The minimum of the red $I_c(H)$ curve at zero field is due to inclusion of the π-junction in the superconducting loop. In the conventional SQUID the same frustrated state exists at an external magnetic field corresponding to half-integer numbers of magnetic flux quanta per cell. Thus, embedding an SFS π phase shifter into a superconducting loop indeed leads to self-biasing of the loop by a spontaneously induced supercurrent.

Complementary DC-SQUIDs realised in this experiment are the basic cells of complementary Josephson junction logic proposed by Terzioglu and Beasley [12]. DC-SQUIDs with SFS π-junction phase shifters and similarly built cells [3] can be also used in Single-Flux-Quantum (SFQ) logic circuits. The SFQ logic circuits enable processing of information in the form of single flux quanta which can be stored in elementary superconducting cells including inductors and Josephson junctions. Dynamically, this information is represented by SFQ voltage pulses [19] having a quantised area $\int V(t)dt = \Phi_0$ and corresponding to the transfer of one flux quantum across a Josephson

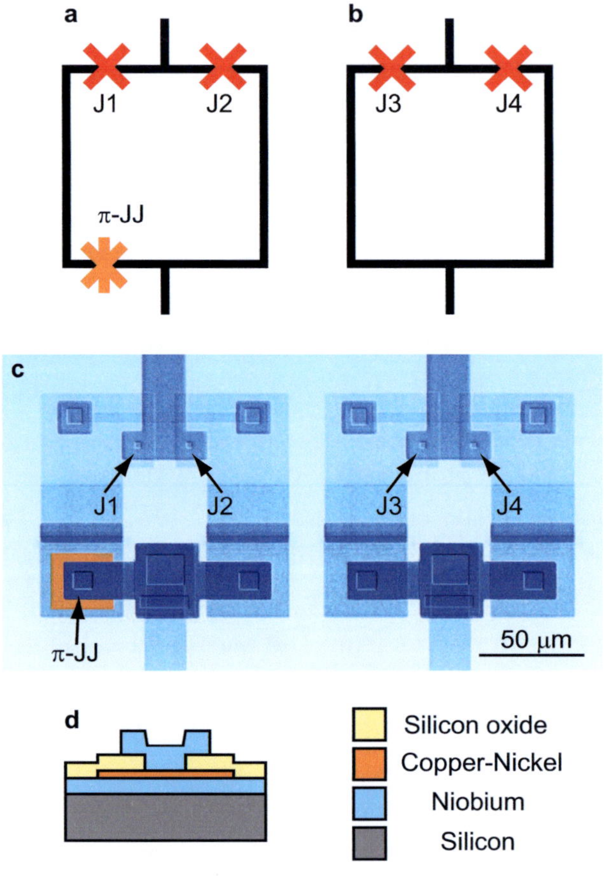

Figure 4.1: Complementary DC-SQUIDs. **a**, Schematic of a complementary DC-SQUID employing two conventional Josephson junctions (red crosses) and a π-junction (orange star). **b**, Schematic of a conventional DC-SQUID used as a reference device. **c**, A SEM micrograph of the fabricated DC-SQUIDs. The ferromagnetic layer is shown in orange. **d**, Schematic cross-section through an SFS π-junction.

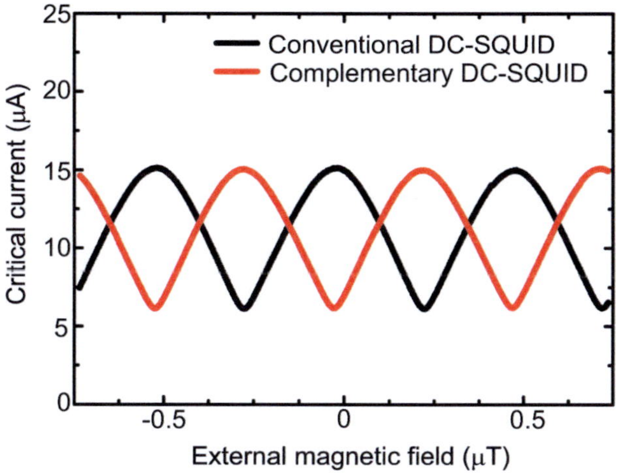

Figure 4.2: Magnetic field dependencies of complementary DC-SQUIDs. Dependencies of the critical currents of the devices shown in Fig. 4.1(c) vs. the applied magnetic field. The red curve related to the π-SQUID is shifted by half a period. Modulation amplitude is limited, since the factor $2LI_c \approx 0.85\ \Phi_0$.

junction. The first SFQ circuits with active π elements were made of high-T_c superconductor ($YBa_2Cu_3O_{7-\delta}$) employing d-wave pairing symmetry combined with conventional low-T_c superconductor (Nb) [20]. Operation of the circuits with the phase shifting element based on frozen flux quanta [21] has been tested earlier in Ref. [22].

4.5 The flux biased phase qubit with an SFS π-shifter

Another attractive application of SFS π-junctions is their use as phase shifters in coherent quantum circuits realising superconducting quantum bits. The answer to the question of whether or not π-junctions can become useful in superconducting circuits designed for quantum computing applications depends on their impact on the coherence properties of the qubits. Potential sources of decoherence introduced by π-junctions can for instance be spin-flips in the ferromagnetic barrier [23], either occurring randomly or being

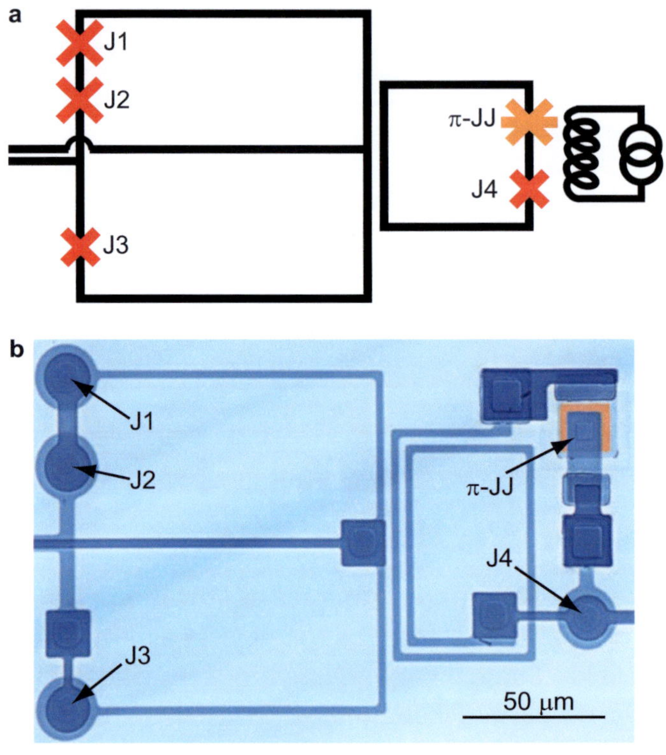

Figure 4.3: Self-biased phase qubit. **a**, Schematic of a phase qubit circuit used to test the decoherence properties of the π-junction. The qubit is realised by the central loop with embedded conventional and π - Josephson junctions. The larger loop to its left is a DC-SQUID for qubit readout. To the right of the qubit is a coupled weakly flux bias coil. **b**, SEM picture of the realised phase qubit employing a π - junction in the qubit loop. The flux bias coil is not shown.

driven by high-frequency currents and fields, as well as the dynamic response of the magnetic domain structure[24]. We address these important coherence issues in another experiment, in which we use an SFS π-junction to self-bias a superconducting phase qubit. We have chosen here a phase qubit [25] rather than a flux qubit [26] due to the simpler fabrication procedure for the former. The results reported below would nevertheless remain fully applicable to flux qubits.

A phase qubit [25] consists of a single Josephson junction embedded in a superconducting loop. It is magnetically biased close to an integer number of flux quanta in the loop. At such a bias, the potential energy of the qubit exhibits an asymmetric double-well potential, whereas two quantised energy eigenvalues of the phase localised inside the shallow well are used as the logical qubit states $|0\rangle$ and $|1\rangle$. Figure 4.3(a) shows a circuit schematic and 4.3(b) a micrograph of the tested sample. Here, a π-junction is connected in series to the phase qubit's tunnel junction. The qubit is controlled by inducing a small-amplitude microwave current in the loop whose frequency is tuned in resonance to the $|0\rangle$ to $|1\rangle$ transition, giving rise to Rabi oscillation of the state population. Reading out the qubit is accomplished by applying a short dc flux pulse to the qubit loop, during which only the excited qubit state may tunnel to the neighbouring potential well. Since this tunneling event entails a flux quantum entering the qubit loop, reading out the qubit is concluded by a measurement of the flux threading the qubit loop by means of an inductively coupled DC-SQUID. Coherent qubit operation is demonstrated by the data reported in Fig. 4.4(a), showing Rabi oscillation of the excited qubit state population probability in dependence on the duration of a resonant microwave pulse. The oscillations exhibit a decay time of about 4 ns, which is a typical value reachable in samples fabricated using similar fabrication processes [27]. To find out whether π-junction does introduce extra decoherence, a conventional phase qubit without a π-junction was fabricated on the same wafer. As shown in Fig. 4.4(b), this reference qubit shows a nearly identical decay time for Rabi oscillations.

Thus we obtained experimentally that the decay time limited by π-junction is significantly larger than 4 ns.

Figure 4.5 shows a dependence of self-biased phase qubit coherent oscillation frequency on amplitude of the resonant microwave drive. The dependence is in an agreement with earlier observations on conventional phase qubits fabricated by similar technological process [28].

The excited state of the self-biased phase qubit exhibits an exponential decay as expected (see Fig. 4.6). The energy relaxation time T_1 is about 5 ns.

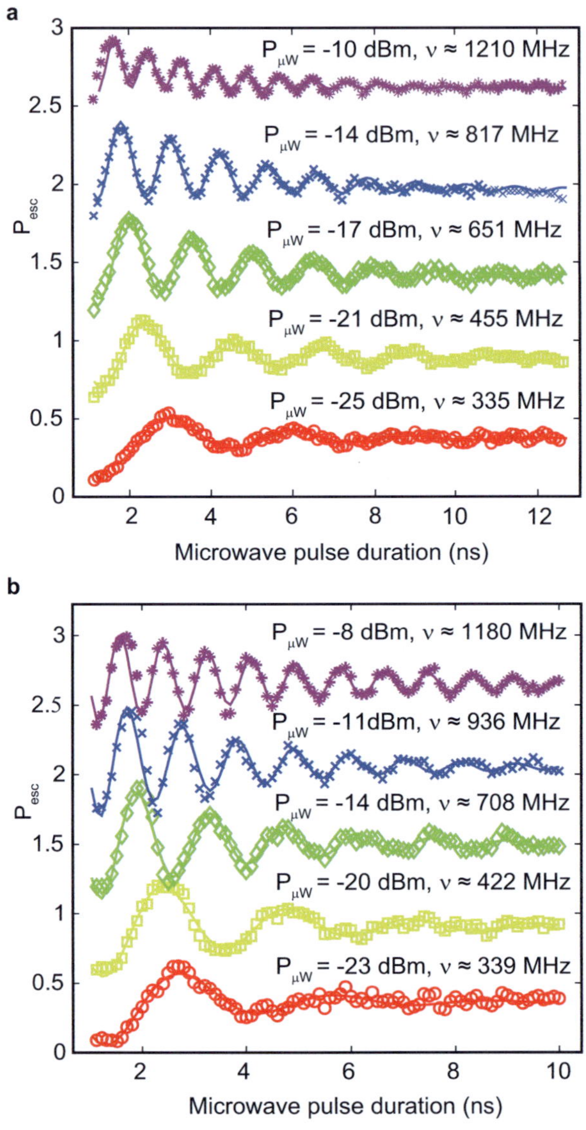

Figure 4.4: Rabi oscillations between the ground and the excited qubit states resulted from resonant microwave driving. a, Observed in the phase qubit with embedded π-junction, and **b**, A conventional phase qubit made on the same wafer as a reference. Each data set was taken using the indicated microwave power as delivered by the generator, giving rise to a change in the coherent oscillation frequency as expected for Rabi oscillation.

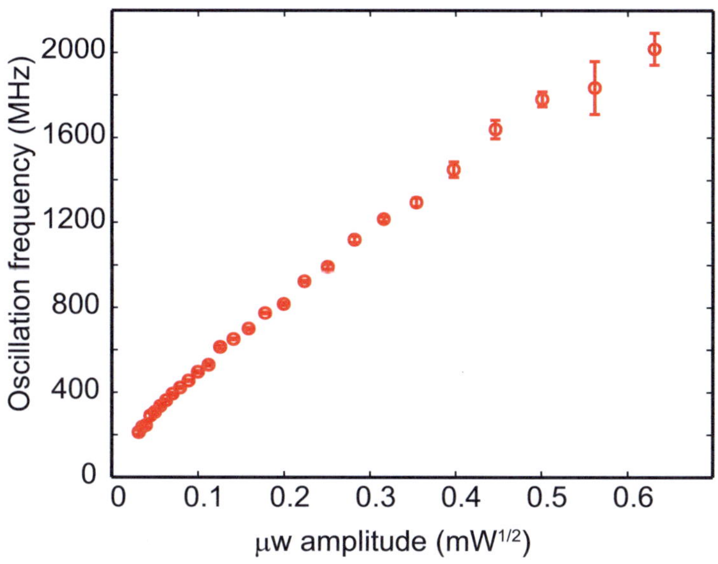

Figure 4.5: Dependence of Rabi oscillation frequency on microwave amplitude for the self-biased phase qubit.

4.6 Estimates of decoherence in π-junctions

We compared the measured decoherence time with the theoretical predictions [29]. We assume here an *overdamped* SFS π-junction having a normal resistance of $R_{N,\pi} \approx 500\ \mu\Omega$ and a critical current $I_{C,\pi} \approx 50\ \mu$A. In our case, the qubit level splitting $\Delta \gg 2eI_{C,\pi}R_{N,\pi}$, where $\Delta \approx h\cdot$ 13.5 GHz, h is the Plank's constant and e is the elementary charge. Here, the energy $2eI_{C,\pi}R_{N,\pi} \approx h\cdot 12\,\text{MHz}$ is associated with characteristic Josephson frequency of our SFS π-junction.

In the paper by Kato, Golubov and Nakamura [29], the following expressions for the effective noise spectrum J_{eff} and relaxation time τ_{relax} were obtained:

$$J_{\text{eff}}(\omega) = \frac{8E_J^2 E_{C,\pi}}{\hbar^3}\left[\frac{\gamma\omega}{\gamma^2\omega^2 + (\omega^2 - \omega_0^2)^2}\right], \tag{4.1}$$

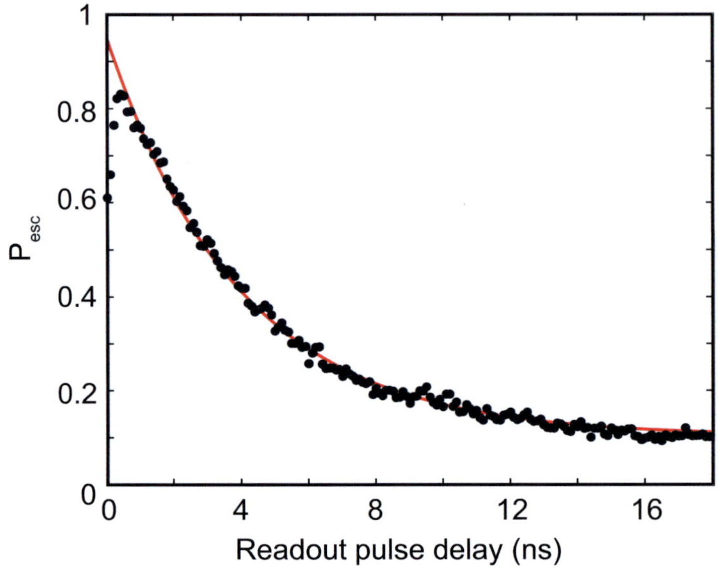

Figure 4.6: Decay of the first excited state of the self-biased phase qubit. $T_1 \approx 5$ ns

where $\gamma = \frac{1}{R_{N,\pi}C_\pi}$, $\omega_0 = \frac{\sqrt{8E_{J,\pi}E_{C,\pi}}}{\hbar}$, $E_{C,\pi} = \frac{e^2}{2C_\pi}$ and

$$\tau_{\text{relax}}^{-1} = 2J_{\text{eff}}(\Delta/\hbar)\coth\left(\frac{\Delta}{2k_BT}\right), \tag{4.2}$$

where Δ is the qubit energy level splitting. These equations were derived using an RCSJ model under an assumption $E_{J,\pi} \gg E_J \gg E_C$, which means the π-junction remains in the superconducting state and is operated in the regime of a well defined phase.

For our case, we should take into account that the capacitance C_π of our SFS junction is very small (less than 10 fF according to our estimate).

Assuming therefore that $\omega \ll \omega_0$, we can write

$$J_{\text{eff}}(\omega) \approx \frac{8E_J^2 E_{C,\pi}}{\hbar^3}\left[\frac{\gamma\omega}{\gamma^2\omega^2+\omega_0^4}\right] = \tag{4.3}$$

$$= \frac{8E_J^2 E_{C,\pi}}{\hbar^3}\left[\frac{\gamma\omega}{\gamma^2\omega^2+64\frac{E_{J,\pi}^2 E_{C,\pi}^2}{\hbar^4}}\right] = \tag{4.4}$$

$$= \frac{4E_J^2 \frac{e^2}{R_{N,\pi}C_\pi^2}\omega}{\frac{\hbar^3}{R_{N,\pi}^2 C_\pi^2}\left[\omega^2+16\frac{E_{J,\pi}^2}{\hbar^4}e^4 R_{N,\pi}^2\right]}. \tag{4.5}$$

Now we can exclude C_π ,

$$J_{\text{eff}}(\omega) \approx \frac{4E_J^2 e^2 R_{N,\pi}\omega}{\hbar^3\left[\omega^2+16\frac{E_{J,\pi}^2}{\hbar^4}e^4 R_{N,\pi}^2\right]}. \tag{4.6}$$

Since $E_J = \frac{I_C\Phi_0}{2\pi} = \frac{I_C\hbar}{2e}$, it allows us to further simplify the latter expression:

$$J_{\text{eff}}(\omega) \approx \frac{4\frac{I_C^2\hbar^2}{4e^2}e^2 R_{N,\pi}\omega}{\hbar^3\left[\omega^2+16\frac{I_{C,\pi}^2\hbar^2}{4e^2}\frac{e^4}{\hbar^4}R_{N,\pi}^2\right]} = \tag{4.7}$$

$$= \frac{I_C^2 R_{N,\pi}\hbar\omega}{\hbar^2\omega^2+4e^2 I_{C,\pi}^2 R_{N,\pi}^2}. \tag{4.8}$$

The qubit relaxation time referred to the frequency $\omega = \Delta/\hbar$ can be calculated as

$$\tau_{\text{relax}} = \left[2J_{\text{eff}}(\Delta/\hbar)\coth\left(\frac{\Delta}{2k_BT}\right)\right]^{-1} \approx \tag{4.9}$$

$$\approx \left[2\left(\frac{I_C^2 R_{N,\pi}\hbar\frac{\Delta}{\hbar}}{\hbar^2\left(\frac{\Delta}{\hbar}\right)^2+4e^2 I_{C,\pi}^2 R_{N,\pi}^2}\right)\coth\left(\frac{\Delta}{2k_BT}\right)\right]^{-1} = \tag{4.10}$$

$$= \left(\frac{\Delta^2+4e^2 I_{C,\pi}^2 R_{N,\pi}^2}{2I_C^2 R_{N,\pi}\Delta}\right)\tanh\left(\frac{\Delta}{2k_BT}\right). \tag{4.11}$$

If the qubit level splitting $\Delta \gg 2eI_{C,\pi}R_{N,\pi}$ we can neglect the second term in the numerator:

$$\tau_{\text{relax}} \approx \left(\frac{\Delta}{2I_C^2 R_{N,\pi}}\right)\tanh\left(\frac{\Delta}{2k_BT}\right) \approx 2\,\text{ns}. \tag{4.12}$$

Temperature T was about 50 mK, thus $\tanh\left(\frac{\Delta}{2k_BT}\right) \approx 1$. Here $\Delta \approx h \cdot 13.5\,\text{GHz}$, $I_C \approx 2\,\mu\text{A}$ is the critical current of the small SIS qubit junction, and $R_{N,\pi} \approx 500\,\mu\Omega$. Thus, in our case of strongly overdamped π-junction, qubit relaxation time is limited by the ohmic losses in π-junction.

Finally we note that, in general, expression (4.12) remains valid without restriction to the case $\omega \ll \omega_0$, under a softer assumption that $\gamma^2\omega^2 \gg (\omega^2 - \omega_0^2)^2$. The qubit relaxation time given by the equation (4.12) should thus provide the lowest possible value corresponding to the worst case scenario.

The estimated value of the energy relaxation time is of the same order as the measured decoherence time of our reference qubit without a SFS π-junction. In fact, the theoretical estimate is valid for the qubit operated as a flux qubit. In the phase qubit regime, the relaxation time is expected to be longer, since the amplitude of the microwave current in the qubit loop is much smaller than I_C, allowing us to conclude that at least on the observable time scale no extra decoherence is introduced by the SFS π phase shifter employed in this circuit and that the decoherence in both qubits is limited by some other mechanism. We note, however, that the expected relaxation time 4.12)can be enhanced by using SFS junctions having a smaller resistance $R_{N,\pi}$.

4.7 Discussion

In contrast to π-junctions based on high-T_c superconductor junctions with d-wave pairing symmetry, SFS elements can have a sufficiently large critical current, so that the desired π phase shift remains reliably fixed during circuit operation. In distinction from phase-shifting loops with frozen magnetic flux [21], the SFS circuits are much more compact and do not require trapping a well-defined integer number flux quanta in their superconducting loops.

As an outlook, a significant reduction in the size of the demonstrated SFS π-phase shifters for digital circuits is readily possible, opening the way to scaling superconducting logic circuits down to sub-μm dimensions [3]. The visualisation of the magnetic structure of our F layer material shows domain sizes smaller than 100 nm. Therefore, we believe that a reduction of the junction planar dimensions down to 300-500 nm is feasible. Furthermore, combining the high-j_C π-junction technology with in-situ grown tunnel barriers [30, 31] may open the way towards active inverter elements which are in great demand for superconducting electronics.

4.8 Summary

We demonstrated here the successful operation of two generic superconducting circuits with embedded π-junction phase shifters. For the studied π-biased phase qubit, we observed Rabi oscillations and compared their coherence time with that of conventional phase qubits fabricated by the same technology. We find no degradation of the coherence time induced by the presence of the π-junction. The demonstrated SFS π-junction phase shifter circuits are readily scalable and compatible with conventional niobium-based superconducting circuit technology.

References

[1] Bulaevskii, L. N., Kuzii, V. V. & Sobyanin, A. A. Superconducting system with weak coupling to the current in the ground state. *JETP Lett.* **25,** 290-294 (1977).

[2] Buzdin, A. I., Bulaevskij, L. N. & Panyukov, S. V. Critical-current oscillations as a function of the exchange field and thickness of the ferromagnetic metal (F) in an S-F-S Josephson junction. *JETP Lett.* **35,** 178-180 (1982).

[3] Ustinov, A. V. & Kaplunenko, V. K. Rapid single-flux quantum logic using π-shifters. *J. Appl. Phys.* **94,** 5405-5407 (2003).

[4] Ioffe, L. B., Geshkenbein, V. B., Feigelman, M. V., Fauchère, A. L. & Blatter, G. Environmentally decoupled sds-wave Josephson junctions for quantum computing. *Nature* **398,** 679-681 (1999).

[5] Blatter, G., Geshkenbein, V. B. & Ioffe, L. B. Design aspects of superconducting-phase quantum bits. *Phys. Rev. B* **63,** 174511 (2001).

[6] Ryazanov, V. V. *et al.* Coupling of two superconductors through a ferromagnet: evidence for a π-junction. *Phys. Rev. Lett.* **86,** 2427-2430 (2001).

[7] Van Harlingen, D. J. Phase-sensitive tests of the symmetry of the pairing state in the high-temperature superconductors—Evidence for $d_{x^2-y^2}$ symmetry. *Rev. Mod. Phys.* **67,** 515-535 (1995).

[8] Baselmans, J. J. A., Morpurgo, A. F., van Wees, B. & Klapwijk, T. M. Reversing the direction of supercurrent in a controllable Josephson junction. *Nature* **397,** 43-45 (1999).

[9] Testa, G. *et al.* Midgap state-based π-junctions for digital applications. *Appl. Phys. Lett.* **85,** 1202-1204 (2004).

[10] Hilgenkamp, H. *et al.* Ordering and manipulation of the magnetic moments in large-scale superconducting π-loop arrays. *Nature* **422,** 50-53 (2003).

[11] Cleuziou, J.-P., Wernsdorfer, W., Bouchiat, V., Ondarcuhu, T. & Monthioux, M. Carbon nanotube superconducting quantum interference device. *Nature Nanotech.* **1,** 53-59 (2006).

[12] Terzioglu, E. & Beasley, M. R. Complementary Josephson junction devices and circuits: a possible new approach to superconducting electronics. *IEEE Trans. Appl. Supercond.* **8,** 48-53 (1998).

[13] Buzdin, A. I. Proximity effects in superconductor-ferromagnet heterostructures. *Rev. Mod. Phys.* **77,** 935-976 (2005).

[14] Frolov, S. M. *et al.* Imaging spontaneous currents in superconducting arrays of π-junctions. *Nature Phys.* **4,** 32-36 (2008).

[15] Ryazanov, V. V., Oboznov, V. A., Veretennikov, A. V. & Rusanov, A. Y. Intrinsically frustrated superconducting array of superconductor-ferromagnet-superconductor π junctions. *Phys. Rev. B* **65,** 020501 (2001).

[16] Frolov, S. M., Van Harlingen, D. J., Oboznov, V. A., Bolginov, V. V. & Ryazanov, V. V. Measurement of the current-phase relation of superconductor/ferromagnet/superconductor π Josephson junctions. *Phys. Rev. B* **70,** 144505 (2004).

[17] Oboznov, V. A., Bolginov, V. V., Feofanov, A. K., Ryazanov, V. V. & Buzdin, A. I. Thickness dependence of the Josephson ground states of

superconductor-ferromagnet-superconductor junctions. *Phys. Rev. Lett.* **96,** 197003 (2006).

[18] Burmistrov, E. V. *et al.* A planar picoamperemeter based on a superconducting quantum interferometer. *J. Commun. Technol. Electr.* **51,** 1319-1324 (2006).

[19] Likharev, K. K. & Semenov, V. K. RSFQ logic/memory family: a new Josephson-junction technology for sub-terahertz-clock-frequency digital systems. *IEEE Trans. Appl. Supercond.* **1,** 3-28 (1991).

[20] Ortlepp, T. *et al.* Flip-flopping fractional flux quanta. *Science* **312,** 1495-1497 (2006).

[21] Majer, J. B., Butcher, J. R. & Mooij, J. E. Simple phase bias for superconducting circuits. *Appl. Phys. Lett.* **80,** 3638-3640 (2002).

[22] Balashov, D. *et al.* Passive phase shifter for superconducting Josephson circuits. *IEEE Trans. Appl. Supercond.* **17,** 142-145 (2007).

[23] Mélin, R. $\sin(2\varphi)$ current-phase relation in SFS junctions with decoherence in the ferromagnet. *Europhys. Lett.* **69,** 121-127 (2005).

[24] Hikino, S., Mori, M., Takahashi, S. & Maekawa, S. Ferromagnetic resonance induced Josephson current in a superconductor/ferromagnet/superconductor junction. *J. Phys. Soc. Jpn.* **77,** 053707 (2008).

[25] Simmonds, R. W. *et al.* Decoherence in Josephson phase qubits from junction resonators. *Phys. Rev. Lett.* **93,** 077003 (2004).

[26] Chiorescu, I., Nakamura, Y., Harmans, C. J. P. M. & Mooij, J. E. Coherent quantum dynamics of a superconducting flux qubit. *Science* **299,** 1869-1871 (2003).

[27] Lisenfeld, J., Lukashenko, A., Ansmann, M., Martinis, J. M. & Ustinov, A. V. Temperature dependence of coherent oscillations in Josephson phase qubits. *Phys. Rev. Lett.* **99,** 170504 (2007).

[28] Lisenfeld, J. *Experiments on superconducting Josephson phase quantum bits.* (PhD thesis, Friedrich-Alexander-Universität Erlangen-Nürnberg, 2008).

[29] Kato, T., Golubov, A. A. & Nakamura, Y. Decoherence in a superconducting flux qubit with a π-junction. *Phys. Rev. B* **76,** 172502 (2007).

[30] Weides, M. *et al.* 0-*pi* Josephson tunnel junctions with ferromagnetic barrier. *Phys. Rev. Lett.* **97,** 247001 (2006).

[31] Bannykh, A. A. *et al.* Josephson tunnel junctions with a strong ferromagnetic interlayer. *Phys. Rev. B* **79,** 054501 (2009).

Summary

This thesis presents experimental work aimed at investigation of superconducting flux qubits that feature an orthogonal control of the qubit potential. These qubits are promising candidates for the implementation of scalable quantum information processing. This approach allows to combine qubit frequency tunability with the optimal coherence at the potential symmetry point.

After general introduction given in Chapter 1, our flux qubit design where the splitting between the two lowest energy levels can be changed *in situ* is introduced in Chapter 2. A topological modification that allows for an orthogonal control over the qubit Hamiltonian is described. We observe a switching of the two-cell flux qubit with a passive π-shifter between two classical states. The switching occurs around zero flux for both symmetric and antisymmetric flux components. That indicates that the proposed "silent" two-cell qubits are insensitive in the linear order to fluctuations caused by the flux control circuitry. Our experiments have thus shown an approach to minimise the coupling between a qubit and its environment while keeping the full control over the circuit.

Chapter 3 presents an experimental investigation of a gradiometer flux qubit with a tunable gap coupled to a lumped-element LC resonator mode in the SQUID circuitry. We found strong coupling between the qubit and the cavity. Vacuum Rabi oscillations are observed for different qubit gaps, including the case where the qubit gap is equal to the resonant frequency of the cavity. This allows the qubit to interact with the cavity at the symmetry point where qubit coherence is optimal, and where the coupling between the two systems is maximal. This experiment opens a way to implement a cavity bus for flux qubits.

In Chapter 4, an experimental implementation of a passive π-shifter based on superconductor-ferromagnet-superconductor Josephson junctions is presented. The successful operation of two generic superconducting circuits with embedded π-junction phase shifters is demonstrated. For the studied π-biased phase qubit, no degradation of the coherence time induced by the presence of the π-junction is observed. The demonstrated SFS π-junction phase shifter circuits are readily scalable and compatible with conventional niobium-based superconducting circuit technology.

The results of the research reported in this thesis show that tunable gap flux qubits have a potential for building quantum registers. Cavities coupled to flux qubits can be used for information storage and transfer between qubits. SFS π-shifters provide a simple approach to bias multi-qubit circuits. π-junction biased flux qubits are expected to be realised soon. A possibility to change the qubit resonance frequency while preserving qubit coherence enables implementation of switchable coupling between qubits and cavities. Thus circuit quantum electrodynamics with flux qubits renders an excellent test-bed to investigate light-matter interaction and to apply the gained knowledge to quantum information science.

Alexey Feofanov
Karlsruhe, November 2010

Zusammenfassung

Das Ziel der hier vorgestellten Arbeit war die experimentelle Untersuchung von supraleitenden Fluss-Qubits, welche als Besonderheit die unabhängige Kontrolle des Qubit-Potenzials erlauben. Diese Qubits haben ein grosses Potential zur Realisierung skalierbarer Strukturen für die Quanteninformationsverarbeitung. Unser Ansatz vereint eine Abstimmbarkeit der Qubit-Resonanzfrequenz mit dem Erhalt der optimalen Kohärenz am Symmetriepunkt des Qubit-Potenzials. Nach einer allgemeinen Einführung im ersten Kapitel wird in Kapitel zwei unser Fluss-Qubit Design vorgestellt, welches eine *in situ* Änderung des Abstands zwischen den beiden niedrigsten Energieniveaus des Qubits erlaubt. Wir beschreiben eine topologische (geometrische) Modifikation des Qubits, die eine unabhängige Kontrolle des Qubit Hammilton-Operators ermöglicht. An einem solchen Zwei-Schleifen Fluss-Qubit mit integriertem passivem π-Phasenschieber konnten wir ein Schalten zwischen zwei klassischen Flusszuständen beobachten. Dieses Schalten trat in der Nähe der Nullpunkte sowohl der symmetrischen als auch der antisymmetrischen Flusskomponente auf. Dies zeigt, dass das vorgeschlagene "stille" zweischleifige Qubit in linearer Ordnung unempfindlich gegenüber Rauschen ist, dass aus dem zur Flusskontrolle notwendigen Schaltungsteil herrührt. Unsere Experimente haben damit eine Möglichkeit gezeigt, die Kopplung zwischen einem Qubit und der Umgebung zu Minimieren, während gleichzeitig die volle Kontrolle gewährt bleibt.

Im dritten Kapitel präsentieren wir experimentelle Untersuchungen an einem voll verstimmbaren gradiometrischen Fluss-Qubit, welches an einen LC-Resonator gekoppelt ist, der von dem SQUID-Detektor gebildet wird. Hierbei beobachteten wir eine starke Kopplung zwischen dem Qubit und dem

Resonator und konnten Vakuum-Rabi-Schwingungen für verschieden Abstände der Energieniveaus messen. Dabei betrachteten wir auch den Spezialfall, bei dem der Abstand der zwei niedrigsten Energieniveaus des Qubits gerade der Resonanzfrequenz des SQUID-Detektors entspricht. Dies ermöglicht eine Qubit-Resonator Wechselwirkung bei optimalen Kohärenzbedingungen am Symmetriepunkt, bei gleichzeitig grösstmöglicher Stärke der Wechselwirkung. Dieses Experiment zeigt daher einen Weg zur Realisierung eines "Bus-Resonators" für Fluss-Qubits auf.

Im vierten Kapitel beschreiben wir schließlich eine experimentelle Realisierung eines passiven π-Phasenschiebers, der auf einem Supraleiter - Ferromagnet - Supraleiter Josephsonkontakt basiert. Wir zeigen die Funktionsfähigkeit von zwei auf gewöhnlichen Josephsonkontakten basierenden Strukturen, die nachträglich mit π-Kontakt Phasenschiebern versehen wurden. An einem durch einen π-Kontakt versorgtem Phasen-Qubit wurde keine Verringerung der Kohärenzzeit durch die Integration des π-Kontakts beobachtet. Die vorgestellten SFS π-Kontakt Phasenschieber sind skalierbar sowie kompatibel zu konventionellen, Niob-basierten Herstellungsprozessen für supraleitende Schaltkreise.

Die in dieser Arbeit vorgestellten Forschungsergebnisse zeigen, dass abstimmbare Fluss-Qubits das Potential zur Realisierung von Quantenregistern haben. An solche Qubits gekoppelte Resonatoren können sowohl als Speicher für Quanteninformation als auch zur Zustandsübertragung zwischen Qubits dienen. SFS π-Phasenschieber bieten eine einfache Möglichkeit, aus mehreren Qubits bestehende Systeme an ihrem Arbeitspunkt zu betreiben. Die erste Realisierung eines Fluss-Qubits mit eingebettetem π-Kontakt ist in naher Zukunft zu erwarten. Die Möglichkeit, die Resonanzfrequenz des Qubits zu ändern und dabei seine Kohärenz zu erhalten, erlaubt die Realisierung einer schaltbaren Kopplung zwischen Resonatoren und Qubits. Fluss-Qubits liefern daher exzellente Voraussetzungen für die weitere Untersuchung der Wechselwirkung zwischen Licht und Materie im Bereich der Quanten-Elektrodynamik, während die dabei gewonnenen Erkenntnisse der Wissenschaft der Quanteninformationsverarbeitung dienen.

Alexey Feofanov

Karlsruhe, November 2010

Acknowledgements

I would like to thank everyone who contribute to this work in any way. In particular I am grateful to Prof. Dr. Alexey Ustinov for being my supervisor. He made it possible for me to join his group and take part in the group research and other group activities including but not limited to ski seminars. Especially I would like to thank Alexey for teaching me climbing and skiing.

I also appreciate the help Dr. Alexandr Lukashenko gave me with handling and organising cryogenic setups. I also thank him for sharing his knowledge about cryogenic equipment and properties of materials at low temperatures. And of course I cannot but admire his stories and songs.

I am grateful to Dr. Jürgen Lisenfeld for his help with organising measurements and for his measurement software. Jürgen, I also thank you for your cooperation when we were sharing a flat in Delft. I thank Tobias Wirth for his help with electronics and microwave equipment. Tobi, I am glad that you were accompanying me throughout my stay in the group. Thank you for your support and nice atmosphere you were creating.

I appreciate Dr. Stefano Poletto supporting our coffee machine. This piece of equipment is vital for the research. I also thank him as well as Dr. Pavel Bushev and Dr. Hannes Rotzinger for manifold discussions. I am also grateful to Susanne Butz and Grigorij Grabovskij for their patience, when they had to attend my exercises class on superconductivity. I thank all the members of the group both former and current for keeping the group up.

I thank Prof. Dr. Hans Mooij and Dr. Kees Harmans for their hospitality and support during my stay in the Quantum Transport group at the Kavli Institute of Nanoscience at Delft University of Technology (QT). Hans and Kees, it was a pleasure to communicate to you and to learn from you.

I appreciate Dr. Arkady Fedorov and Dr. Pol Forn-Díaz working together with me. I think the success of our experiments was founded on this collaboration. I thank Dr. Pieter de Groot and Reinier Heeres for writing QT Lab measurement environment. Pieter, thank you for helping me with Python and measurement scripts. I admire the electronics designed by Raymond Schouten. This electronics made our measurements possible. I thank the whole Flux Qubit Team for useful discussions and the whole QT for a nice atmosphere.

I am also grateful to all my co-authors for their contribution. I hope to continue our collaboration in the future.

Curriculum Vitae

Alexey Feofanov

Oct 19, 1982	Born in Zaporozhye, the USSR.
1989-1990	Secondary School N74, Zaporozhye.
1990-1996	Secondary School N51, Zaporozhye.
1996-1999	Lycée "Logos", Zaporozhye.
1999-2005	M. Sc. Applied Physics and Mathematics, Moscow Institute of Physics and Technology. Graduate research in the Laboratory for Superconductivity, Institute of Solid State Physics, Russian Academy of Sciences under supervision of Prof. Dr. V. V. Ryazanov.
2005-2010	Ph.D. research at the University of Erlangen-Nürnberg and Karlsruhe Institute of Technology (KIT) under supervision of Prof. Dr. A. V. Ustinov. Subject: Experiments on flux qubits with π-shifters.
2009-2010	A visit to Delft University of Technology, research in the Quantum Transport qroup under supervision of Prof. Dr. Ir. J. E. Mooij and Dr. C. J. P. M. Harmans.

List of publications

1. Strong coupling of a quantum oscillator to a flux qubit at its symmetry point.
   A. Fedorov, A. K. Feofanov, P. Macha, P. Forn-Díaz, C. J. P. M. Harmans, and J. E. Mooij.
   Phys. Rev. Lett. **105,** 060503 (2010).

2. Implementation of superconductor-ferromagnet-superconductor pi-shifters in superconducting digital and quantum circuits.
   A. K. Feofanov, V. A. Oboznov, V. V. Bol'ginov, J. Lisenfeld, S. Poletto, V. V. Ryazanov, A. N. Rossolenko, M. Khabipov, D. Balashov, A. B. Zorin, P. N. Dmitriev, V. P. Koshelets, and A. V. Ustinov.
   Nature Phys. **6,** 593 (2010).

3. Static and dynamic properties of 0, pi, and 0-pi ferromagnetic Josephson tunnel junctions.
   J. Pfeiffer, M. Kemmler, D. Koelle, R. Kleiner, E. Goldobin, M. Weides, A. K. Feofanov, J. Lisenfeld, and A. V. Ustinov.
   Phys. Rev. B **77,** 214506 (2008).

4. Thickness dependence of the Josephson ground states of superconductor-ferromagnet-superconductor junctions.
   V. A. Oboznov, V. V. Bol'ginov, A. K. Feofanov, V. V. Ryazanov, and A. I. Buzdin.
   Phys. Rev. Lett. **96,** 197003 (2006).

5. Superconductor-ferromagnet-superconductor pi-junctions.
   V. V. Ryazanov, V. A. Oboznov, A. S. Prokofiev, V. V. Bolginov, and A. K. Feofanov.
   J. Low Temp. Phys. **136,** 385 (2004).

6. Superconducting currents through a ferromagnet. Phase inversion in structures with Josephson pi-junctions.

V. V. Ryazanov, V. A. Oboznov, V. V. Bol'ginov, A. S. Prokof'ev, and A. K. Feofanov.
Usp. Fiz. Nauk (Physics-Uspekhi) **174,** 795 (2004).